Chemistry Cro
Volume 1

Chemistry Crosswords
Volume 1

Compiled by Norman H Keir

RSC Publishing

The molecule on the front cover is reproduced with the kind permission of Pierre Braunstein *et al.* from *Chem. Commun.*, 2005, **21**, 2660.

ISBN 0-85404-689-5
ISSN 1749-6667

A catalogue record for this book is available from the British Library

Published by The Royal Society of Chemistry,
Thomas Graham House, Science Park, Milton Road,
Cambridge CB4 0WF, UK

Registered Charity Number 207890

For further information see our web site at www.rsc.org

Printed by Henry Ling Ltd, Dorchester, Dorset, UK

Introduction

Welcome to the RSC's first chemistry crossword compendium. We have selected 70 crosswords that appeared in Chemistry World's predecessor Chemistry in Britain during the 1980s. Each puzzle is chemistry-related and hopefully you'll find them to be the right mix of challenging and entertaining. If you get stuck, solutions to all the puzzles are included in the back of the book.

Chemistry World and Chemistry in Britain have been running crosswords for the last two decades. To this day, their popularity remains undiminished. The ones in this book were compiled by the late Norman H Keir, who provided the magazines with crosswords for many years.

So, probe the depths of your chemistry knowledge, pit your wits against the skills of the compiler and solve the cryptic clues. Good Luck!

Karen Harries-Rees
Editor, Chemistry World

Crossword no 1

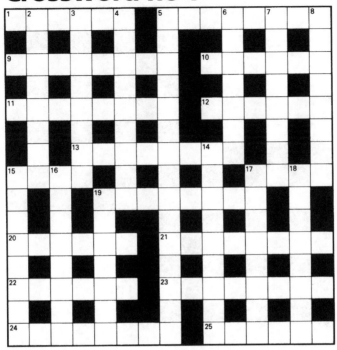

Across

1. Heavy metal's A1 performance in Birmingham (6).
5. Coordinate links that lead to rings, the scale being absurd (8).
9. Device for measuring pX in remote fluid (3,5).
10. Pelt a novice inside a salt compound (6).
11. Constant: a luminous element is enclosed (8).
12. Excavates about 10 jades (6).
13. Leaving out nothing it returns to the start of time in old China (8).
15. Seconds are able to analyse metrically (4).
17. Apprehends those, especially the donor (4).
19. What many 16 often do, sounds outlandish (5,3).
20. Brings forth an elephant or a whale (6).
21. Destroyed by bombing—made it so unrecognisable (8).
22. Student initially tries, is ploughed and tries again (6).
23. Bends having divers applications but mainly cosmetic (8).
24. An amide of this acid could have financed many a trip (8).
25. Moulded, detailed redesign for self-contained unit (6).

Down

2. Capable of electrophilic substitution, like a rose by any other name (8).
3. Moving description of, for example, subatomic particles (2,6).
4. Assemble steel mail to create film (9).
5. Charm too vibrant and vivid for this kind of work done by some 16 (15).
6. Idle nag could make it—but is hardly likely to be (7).
7. Exit Dior dressed—could go with chromium, for example (8).
8. Big lids—restore, lads (8).
14. Cannibalise, Mongolise—a new word (9).
15. Ghostly lines for example (8).
16. New slant, say, for headshrinkers (8).
17. Naked priest—dip could be the outcome, but I must be left out of it (8).
18. Accidental—put across in ceaseless surroundings (8).
19. Confused where teetotalk is one who excites (7).

Crossword no 2

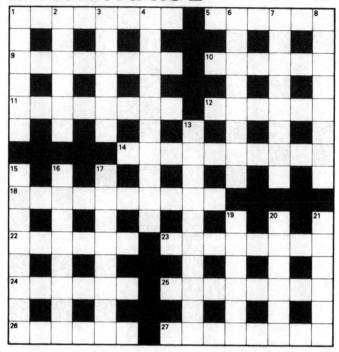

Across

1. and 15. *Down* A kind of agitation—Lord George's tendency? (8,8).
5. Ten? No—mocked by fate (6).
9. Fell clue can provide light (4,4).
10. Trivalent, five not in charge (6).
11. Weather, *eg* three and nineteen (8).
12. Triple negative hydro-formylates to decanol (6).
14. Do reaction components combine harmoniously? Yes (10).
18. Ground rice proved too dear (10).
22. Activate former name (6).
23. In conclusion doers regrouped and expressed approbation (8).
24. Six-footer stands attention before his lordship (6).
25. Unsympathetic object of

Down

1. A dull chap can be useful in electrochemistry (6).
2. In the *Oxford English Dictionary* you'll find the Turkish governor complied with (6).
3. One of 11, 20 to the dollar (6).
4. Descriptive of 11 that have more than one form (10).
6. Develop to richer style of expression (8).
7. Scandinavian in manner so uncouth (8).
8. Pedals around North East to get hydrocarbons (8).
13. What the world needs now—a change would yield useful work (4,6).
15. See 1 *Across*.
16. Of one of 11–80; almost sprightly (8).
17. First pie thrown by stroppy person (8).

Crossword no 3

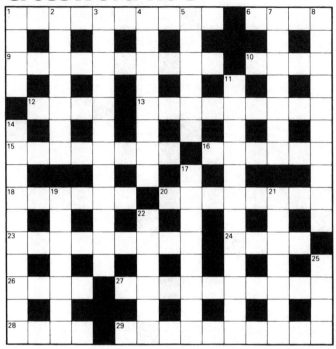

Across

1. Deputy set up, in being replaced by assistant editor (10).
6. Feign in Shakespeare's Hamlet (4).
9. More than one 12 across with engineers *etc* inside; they run on US lines (10).
10. Revise time of flowing, ebbing (4).
12. Nightlight reflected in 28 across (4).
13. Chemical upbuilding needs a pinch o' slim ingredients (9).
15. Ghostly lines for example (8).
16. A shooting 12 across that could go on to become a weather man (6).
18. Produce of slicer can be found in a shrine (6).
20. Harmonise soon—sour sounding (8).
23. A farm rent collector, she might make a fine catch (9).
24. Cupid is touchy when he darts back (4).
26. In measuring length it may have Big Bens all around (4).
27. Emit in source, depositing *in vacuo* (10).
28. 12 across making a comeback—rubbish! (4)
29. Short recipe: sops and seeps—blend to give favourable impression (10).

Down

1. Band of crack troops at start of hostilities (4).
2. Uranium in a better mixture can be used volumetrically (7).
3. One who speculates and the object of mining here in France (in the object of sunbathing) (12).
4. Displaying touch, these works can make a Scot act (8).
5. Unusual dearth in Jenny's output (6).
7. Compound forms a skin round dry mixture (7).
8. Isomerism—surprisingly simmer meat (10).
11. Individuals sign in without feeling of isolation (12).
14. He begins with confusion of 12 across, makes predictions and may be received with 28 across from scientists (10).
17. Poets oft can show the trend which by no tyre was made (8).
19. Ten-line whip-up is emollient (7).
21. A change in zone is nothing; treat with alleged component of fresh air (7).
22. Pelt often found in association with salt (6).
25. Sage reappraisal matures (4).

Crossword no 4

Across
1. Rock bottom, tail first (6).
4. $H_3 PO_4$ is . . . like a milkmaid's stool? (8).
10. Carotenoid, for example, returns Marshal in punt (7).
11. One of a girl's best friends is a card (6).
12. From them you can get music, cart and horses (10).
13. Up-to-date equipment (4).
15. Anonymous and sort of mundane (7).
17. Dosed with I_2 mixture (7).
19. *New Testament* parables about gifts (7).
21. Endless unnecessary irritations (8).
23. Solomon, for example, gets stuffed with onions (4).
24. Trial, if not travestied, can lead to end of suspension (10).
27. All about hydrogen-contaminated CO_2 causing inebriation (7).
28. Ancient galley having three engineers (7).
29. Deranged, pointless, dissolute—can be a lonely place (8).
30. Utensils give abstainer inner comfort (6).

Down
1. Spin-off sounds like what the customers do (9)..
2. Given 50 coins one can obtain the makings of chips (7).
3. Men involved in Linear B decipherment gave name to colour test (10).
5. Given clothing allowance, assistant comes up shining (9).
6. Timber associated with 5 (4).
7. Exhibiters give headstart to broadcasters (7).
8. Evergreen comic raced (5).
9. Trial reagent match (4).
14. This flows with excitement in lined arena (10).
16. Fell about, yet separated (9).
18. Not a fact inside conducive to purify (9).
20. It's reasonable to find a solider in a pub (7).
22. Hangs about—its role needs sorting out (7).
23. Flatters ageless 23 across senior citizens (5).
25. Adds little drinks (4).
26. Snap showing play of colours (4).

Crossword no 5

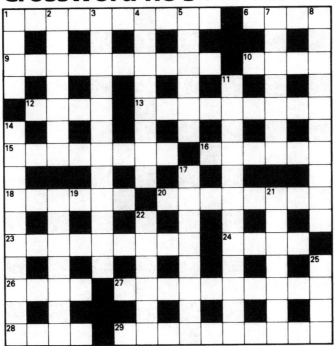

Across

1. Narcotic alkaloid—and no label required (10).
6. Drive before the wind, spank north of the Border (4).
9. Metaphoric mutation acting both ways (10).
10. She's employed in lasers (4).
12. Superb shop redevelopment (4).
13. Granada is there and copper returns among others (9).
15. Divider of angles, twice beheaded Trojan hero (8).
16. Bird lies about will: incomplete and dimly illuminated (6).
18. Old model: three Rs suffering 14 by yeomen half returning (6).
20. If head is left out immersion may be wrong word (8).
23. It's sure a help, unravelled by Willard Gibbs (5, 4).
24. Old coin employed in Siam (4).
26. XV march out of step inside (4).
27. Knock, knock! Not when I follow the lead (10).
28. School return could be a fiver (4).
29. TNT, a phenol derivative, was an event at LA (10).

Down

1. Carry a shaggy carnivore (4).
2. Particles; each is 0.01 Drachma (7).
3. Obtainable from chemists: hose at a price (12).
4. Worker producing drama is over the hill (8).
5. Nymphs produce their first top-class advertisements (6).
7. Assembly sounds like what it may offer (7).
8. Where one gets dyes *etc* sounds like a desiccated old zither (10).
11. Immodest person acting on behalf of another produces gas (7, 5).
14. By this process S and P transform abortion (10).
17. A thief revealed in stream and desert (5, 3).
19. Chemical absent in corrupt Argentine (7).
21. Terpene derived from hot lemon—nil residue (7).
22. Unsat from relocation of NW island to NE (6).
25. Devise scheme (4).

Crossword no 6

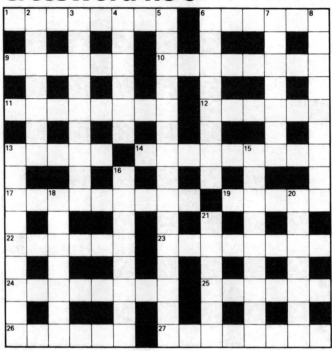

Across

1. Platitudes from newly weds admitting distinction (8).
6. Method is messy to start, very messy (6).
9. Quantity of ale recalls mistake in laboratory (6).
10. Particle causes no tercel to flutter (8).
11. Salt at sea at Easter (8).
12. Recreatively stifle its reflexive form (6).
13. Tittle-tattle is penny tear-jerker (5).
14. Chars sculptured by masseurs (9).
17. I cause precipitation in a remarkable, unskilful process (9).
19. Strengthens partnerships (5).
22. Hold forth about an indefinite number elaborately decorated (6).
23. Speaking in ancient dialect resulting in minute charges (8).
24. Act IV is the setting for one advancing the action (8)
25. Joke on life got from cracker (6).
26. Put up with last obsolete practice (6).
27. Dressed in grey: an intense desire (8).

Down

2. Transformation occurs herein; work of creator (7).
3. Carmen, apt arrangement—this, one can hum (9).
4. A droll, bouncing buck (6).
5. Mercy! So it's there in a new orbit for study of atomic dispositions in space (15).
6. Plausibly deceptive debt memo seen in short glasses (8).
7. A couple of points peter out over $C_{10}H_{16}$ (7).
8. Passenger lists making faint mess (9).
13. Time consumed making salt (9).
15. Be silent—I can translate Handbuch (9).
16. It's a resin or plastic. Like a bet? (8).
18. Editing reprint, reached fire point (7).
20. Remark occurrence of reference (7).
21. Groan! A silly goat! (6).

Crossword no 7

Across

4. Woe is me, initially! I'm in one down (8).
8. Fuzz element (6).
9. In chaste variation of a slender type (8).
10. Flowing out of us—fine! Perhaps (8).
11. I am at Olympia exhibiting a banger (6).
12. Somehow we part at a device that can provide 10 (8).
13. Single comprehensive survey as soon as arriving at an end (4-4).
16. Mix nitre into a hard dry indehiscent fruit. It's nourishing (8).
19. In more ways than one this is the true name for having mathematical nous (8).
21. Managed division of police. That's rank (6).
23. Drunken man involved in hotel affray; that's what comes from alcohol (8).
24. Thoroughly search after spiteful woman, a subterranean excavation (8).
25. Not one painter is resolved to make a hole in the head (6).
26. Return of disorder in porter, a man of steel (8).

Down

1. Two beasts (one possibly dressed as the other) in a native compound (7).
2. Conifer makes one smart with much of great heat (6-3).
3. He performs in dire strait (6).
4. Fraction of current carried. Does it deaden ecstasy? (9,6).
5. Young blood leaning on a half door gets the bird (8).
6. Strangely can set no time for part of an opera (5).
7. Part of a serial record is a poem (7).
14. Caught up with across the margin and seized (9).
15. Chief opportunity afforded by uninterrupted span below ceiling (8).
17. Salt given by a nature lunatic (7).
18. Turkish seat (7).
20. Terriers in dumb change (6).
22. Fish cleaners (5).

Crossword no 8

Across
1. Bright green mineral lets blood flow (6).
5. Salt discharge of uranium battery (8).
9. Oxide and medicinal carbonate in a games assembly (8).
10. A longing was in concord (6).
11. Player's story about tailored suit (8).
12. Heavenly source is in parlous confusion, nothing less (6).
13. Standing nag broken, instant (8).
15. Fuse carried by camel train (4).
17. Ruminants in snug refit (4).
19. Bleeding spots are pollen receptors (8).
20. Area inflamed by herpes (6).
21. Complex substance: it is found in a variety of lichen—and in other living matter (8).
22. Inferior cleric; the scoundrel took food (6).
23. Built her a new structure possibly employed by 22 (8).
24. A product of incandescence can get mother in confusion (8).
25. Walks disaster is missing disaster (6).

Down
2. First little letter used for sounding (4–4).
3. Bus turns out of control in a blaze of light (8).
4. Attacker as ship-worker (9).
5. Ting-a-ling! Stable explosion producer (8,7).
6. Gentile returning injured offers dairy product (7).
7. Dislike a translation (8).
8. New dress one writes on the back of (8).
14. Insignificant and by no means properly called a statement (2–7).
15. Broken mascot is closely related to ... (8).
16. ... Substance the grail is made of (8).
17. Unravel tangle in rope (8).
18. Hungry, internally unwell and at least partly empty (8).
19. Art must change for bed (7).

Crossword no 9

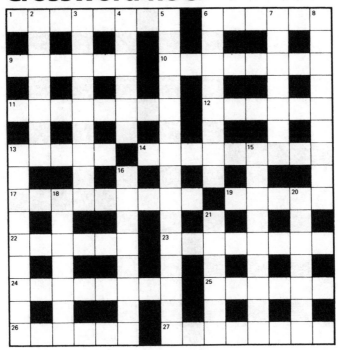

Across

1. I'm live, back in grave; I'm set to go off (4–4).
6. What energy comes in is most of the airline (6).
9. Deep blue and almost hairless in bed (6).
10. Where to find what 26 often uses on his back-end (8).
11. Brightness of nice RADA production (8).
12. Transaction about good book relating to the means of mastication (6).
13. For instance, utter proof (5).
14. Tiny greeting applied to ovens and spectroscopy? (9).
17. Lady gets upset about composition unfavourably (9).
19. L (but not for leather) (5).
22. New angle a lead could be obtained from (6).
23. Take nothing from a storage vessel to provide means of measuring what is removed (3–5).
24. Cyrogenics development company abandoned working together (8).
25. 19 avert disaster to make journey (6).
26. Billiards player is in a cute configuration (6).
27. Passage about disguised lion found in sheep's clothing (8).

Down

2. Is nothing except south running through places under the same pressure (7).
3. Police are up before tea is stirred to uproot (9).
4. Concoct an energy source of motor power (6).
5. React strangely in physiological setting concerned with micro-organisms (15).
6. It's a dilemma—and also right in key (8).
7. Sleep hath confused the spirit (7).
8. Offensive weapons give blood vessel an unfavourable content (9).
13. A nice slag heap can remove pain (9).
15. Policy is bloodless subjugation of student residence (9).
16. Delicacy is French—the goods are inside (8).
18. Drapery combining power (7).
20. The answer could be neither—where else? (7).
21. One was located in dye precursor (6).

Crossword no 10

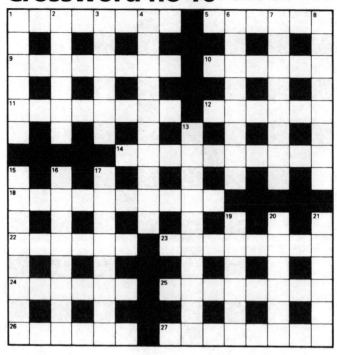

Across

1. His radiation is not green, though he could be (8).
5. Compound in which 100 missed cycling ('going for a spin') (6).
9. Nothing found by police in bail fiasco is lethal (8).
10. Can supremo put back spirit (6).
11. Leave these shores consumed by ghastly grime (8).
12. One of the East-Enders leaves their street for want of a drink (6).
14. Antiseptic for final treatment involving six after car smash (10).
18. New production of a London 'hit' could enhance applause (4–6).
22. First of batsmen in test is characteristic of the breed (6).
23. Describing non-sensual love, one replaces ring of desirable metal (8).
24. A couple of interjections are about satisfactory for a water-pipe (6).
25. Bird fancier has a way with what makes one stir (8).
26. Poet might have this timed with this (6).
27. Murder an organisation for a pit location (8).

Down

1. A spell at cards for the masseur (6).
2. Amorous — can make love in a trice (6).
3. The TA, without hesitation, are men of endeavour (6).
4. The power to recover from depression in the last city one dwells in (10).
6. Readily dispersed, rough, hilly and mostly open (8).
7. Invent novel tin cover (8).
8. One bed in nine is found in 24 (8).
13. Vitamin component synthesised by initially fearless rival boffin (10).
15. Margaret can raise the roof (8).
16. Element is almost a contradiction in law (8).
17. A tendency to split thought by many to be 2 (8).
19. Maybe faster rake (6).
20. Croup, *eg*, puts the soldier in the money (6).
21. Reduction in total gave protection to Romans (6).

Crossword no 11

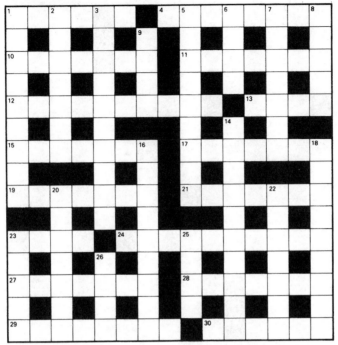

Across

1. Basis of all that is 7 is French for good (6).
4. Reduce to 1 across with fuel, a form of 1 across (8).
10. Kentish town site for banger, south having gone west (7).
11. Through gamut a gentleman runs, an agent of change (7).
12. A duly woven tissue is without acknowledgement (10).
13. Move, by slow degrees, a low-lying meadow beside a river (4).
15. Sketch abroad a service of ships, for instance (7).
17. I have no time for gout ache—get the picture? (7).
19. One of a girl's best friends is a form of 1 across (7).
21. Abridgement of Poe distributed in time warp (7).
23. 23 down for nitroglycerine taken first out of holy-water vessel (4).
24. Saccharin, for example, can be stirred into genuine stew, but not up front (10).
27. Dye precursor—nothing I found in Scots one (7).
28. Betroth sweetheart so prim, but lively (7).
29. A variety of 1 across is a great success in a kind of shot (8).
30. Since bird is behind (6).

Down

1. Decomposed oil culled from this thermoplastic (9).
2. Beaming circular measure takes time (6).
3. Egg layer: it's nothing for half a dozen to perch in humble surroundings (10).
5. What corresponds is colour— and, O, what miserable gloom there is within (9).
6. Esteem ... and scold (4).
7. Containing 1 across and can make one groan with 1 across (7).
8. Throw oneself venturesomely into some activity without a meal (5).
9. Determinant of characteristics is cloth, by the sound of it (4).
14. Yielding gold—the members are furious (10).
16. Demon went bonkers for what is settled (9).
18. Unfading and always vigorous (9).
20. Wild animal absorbs universal oxide (7).
22. Oxford's first team, encountering new side, get rusty (7).
23. Vituperate in terms not accepted for dignified use (5).
25. Catch sight of a member of the KGB? (4).
26. Significant meaning found under orange-peel (4).

Crossword no 12

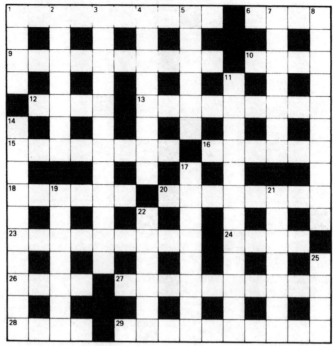

Across

1. Strangely I dreamt it's a consequence of exposure to 4 down (10).
6. Room north of the Border after a border (4).
9. Last letter has nothing written badly about one—despite opposing charges I'm neutral (10).
10. Employer's sure to be organised (4).
12. Without ratio in mathematics (4).
13. Believer taking censure to heart offers treatment (9).
15. I am proof and, in chaos, am urgent (8).
16. Crowd moving north gets to outskirt of Glasgow (6).
18. Getting Scots groups to accept the English makes things look brighter (6).
20. Treat as less important than it is, the opposite of fool (4,4).
23. Try on a shirt—ignore the English, I'm in Britain here (9).
24. Among diamonds, half cut and rearranged (4).
26. The end of life, nothing more, in an age (4).
27. Product of living and it's in meal to be processed (10).
28. What's required sounds like work (4).
29. Having neither strain nor emphasis (10).

Down

1. Bewilderment sounds as if it might last a while (4).
2. Bringing up sound as a bell; an article is included (7).
3. Worker at the steering gear in credit, well almost, deals with worms (12).
4. I storm about most of ceremony, possibly causing 1 across (8).
5. I could be *cis* or *trans*, for example; a little is revealed in infrared (6).
7. The wild one has a celebration. That's chivalry (7).
8. Great stone fashioned can produce monsters (10).
11. Might the initial royal charms produce 19 down (12)?
14. Beware the series that has graced many a corporation (5,5).
17. Behave amorously with spouse for boon companion (8).
19. I rise to a great height, by the sound of it. Not a pretty sight (7).
21. Result of playing violin (a note) can become serpentine (7).
22. Way shortly indistinguishable from holy man (6).
25. The dregs or else otherwise (4).

Crossword no 13

Across

1. Entire flightiness with the wings clipped is unlucky (10).
9. One of the sugars—several do service as source (6).
10. Fuel emerging from pulverised sealing ring (8).
11. Most of class is indeed lowered in reputation (8).
12. Dividing direction of entry (4).
13. Rouse excessively—done by former situation by the sound of it (10).
15. Discount turbulence after loss of direction in the Channel (7).
17. Elementary textbooks can ignite 12 down for example (7).
20. Spot of conviviality, of course, is not far from 1 across (10).
21. Daily, missing a change, is not working (4).
23. Philosopher with a notion—and the inclination (8).
25. After a redesign a lounge results in something similar (8).
26. Give up adorning sweetheart with new rings (6).
27. Fleet seems the makings of pride (4,6).

Down

2. Big push north for God's own country (6).
3. Diamond shaped, battered hood with brim (8).
4. In periodate transformation iodine replaces a mineral (10).
5. Compass pointers are almost unnecessary (7).
6. Suspension would, in church, be evolution (4).
7. A source of wealth, or what belongs to me (8).
8. 5 down cunning is without compulsion (10).
12. For an arsonist it's nice in a dry mixture (10).
14. Trembling with broken heart, friend has lost his right (10).
16. Absurdity is nothing in every which way but west (8).
18. Unlike most vessels this one is no she, by the sound of it (4,4).
19. Place in pew (7).
22. Vegetable, for example, consumed by awkward mule (6).
24. Endless, undisciplined noise is the solution—and possibly in it (4).

Crossword no 14

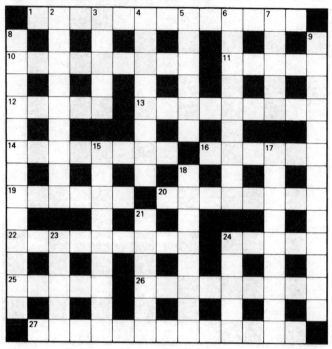

Across

1. High spirits induced by Seidlitz powder (13).
10. Weight initially not tared by doughboy returning smothered in whipped cream (9).
11. Bring back English short story to generate excitement (5).
12. I can raise the dough? Yes, at a new establishment (5).
13. Brutalised molester needs a stretcher (9).
14. Permanganate, perhaps, has two gold coated sides, one in disarray (8).
16. Capers, capers—a long, long way (6).
19. Diamonds: a case for the cooler (6).
20. Aaron's rod, for example—an aniline B derivative (8).
22. Return on cruiser is turbulent (9).
24. Curse says we are where it can be found (5).
25. Preserve and put in dock after dismantling (3,2).
26. Puritan chief follows course which returns upon itself (9).
27. As a mirror is, 'aged' after the tenderest feelings (13).

Down

2. Day-dream: AA participating in development of fitness (9).
3. Be the first of solvers in the way out (5).
4. Cross borne by English voters tossed in maelstrom—more than one in fact (8).
5. Uncle, one of us, has vessel, a small boat (6).
6. I make an effort by ear, the result of reading the small print (9).
7. Mad March quantum number (5).
8. A lament: 'Behold the fool annoy'. It might well end up in the fabric (13).
9. Break the rod—article which, with carbon, can control fire (13).
15. The same in any direction, is nothing with what could be cancer (9).
17. Pivot by agreement, one hears, is squint (6–3).
18. The same duration as 20 or four times as often (8).
21. Tree has most of hard outer coating around it (6).
23. Animal is shy whelp, though not quiet (5).
24. Novice enters team in sideway move (5).

Crossword no 15

Across

1. Deprived of society, recluse displaying acerbity (6).
5. Cleanse as in a mark of distinction (8).
9. A rough centre-cut Pyrex casting used in photographic process (10).
10. Go out regularly with a fruit (4).
11. It's heavenly here in long delayed return (8).
12. Tolerate application? (6).
13. Swelling ebullition (4).
15. Leadership can change in conduct (8).
18. Binding proclamations about day's short duration (8).
19. Pluck and influence (4).
21. I feel important, so I get an facelift (6).
23. Diamond-bearing region of Antarctica? (3–5).
25. Vehicle duty, an irrational quantity (4).
26. Loudspeaker, for example, displayed in current ads (10).
27. Vehicle for final journey by engineers in train (8).
28. Assess a female dolt? (3–3).

Down

2. Swindle—talk about its ending (5).
3. Worked in the garden, equipped as a spur is at the front (9).
4. Badly crated and transported (6).
5. I'm fireproof; my make-up is dreadful, eg politics and death! (15).
6. Alcohol! Golly, a clear sign in an alternative sort of way! (8).
7. Fixed—a thousand left over (5).
8. Using sarcasm, I cast liar out (9).
14. Poem about anger, a disturbance arising from drink (9).
16. Current—illuminated with due change in abundance (9).
17. A foot-slogger, a spud by the sound of it, can stir things up (8).
20. What one comes to on recognition of the facts (6).
22. Compound contribution to the aim I desire (5).
24. Emblematically Cymric, back to front they are smooth (5).

Crossword no 16

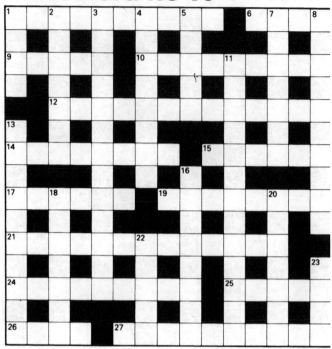

Across

1. Trivial toe disorder—treat with sulphate (10).
6. Newts found in dechlorinated fissures (4).
9. Davy Crocket historical? A mound conceals where he fell (5).
10. Cigar returned by everyone taking 40 winks after tipsy tea (9).
12. Separation of part allowance without resistance (13).
14. Two freshwater fish, the first topped and tailed, intrude (8).
15. Appearance presented as a couple of gallons, say, tea initially (6).
17. Flow regulators save 55 at sea (6).
19. Perhaps George or Andrew has impulse concerning fish (8).
21. Venomous wet slipper (5–8).
24. Jerking hag in ring (9).
25. Where to find sleepers and flowers (2,3).
26. Animal returns for grass (4).
27. Regressive revolutionary campanologists display weapons (10)

Down

1. Meat needlessly and awkwardly leavened (4).
2. Vehicles' disgraceful trading (7).
3. Tribasic and nothing more permutated at the same pressure (13).
4. Lick and spit can strangely brighten one's smile (8).
5. Sounds like a tea dance (5).
7. Rock, rock the last of the dance itself (7).
8. A gnat sting could be getting foul (10).
11. 16's son and I are involved in infringement (13).
13. Reaction moderator is hardly light ale (5,5).
16. Floating tealeaf is a foreigner (8).
18. Wild cattle: one found in network (7).
20. Distinguish a couple of points of high birth (7).
22. Nothing stirring, she is an emblem of peace (5).
23. Poetry does transform (4).

Crossword no 17

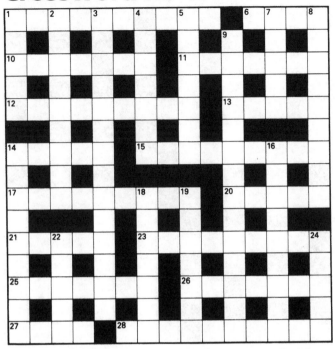

Across
1. Like last month's Chemistry in Britain, say, a thing past use (4-6).
6. Like someone found in torpid leisure (4).
10. Told by dead-centred revolutionary (7).
11. Right in place for birds that is a social circle (7).
12. He coins new names for lotions, for example, perhaps (9).
13. Summing up of penniless military policeman (5).
14. Priest stands before the heartless chosen few (5).
15. Accelerate fibre production for 14 down (9).
17. Doctor's patients put down for selection for member of Brains Trust (9).
20. Rat is possibly one in flight (5).
21. Louse could be a dark-complexioned person (5).
23. Sanguine people make a choice one obscures (9).
25. I'm after nothing — that's laying it on a bit thick (7).
26. Sharp one doubly disguises craft (7).
27. Send back to front for what the prudent would like to meet (4).
28. Slides one into another for taking the long view (10).

Down
1. I'm right in favour, occupying fifth place on the table (5).
2. No colloid could possibly be a form of 3 (9).
3. 2 and 15 for example (14).
4. Hardy crowd, extravagantly exuberant, make a noise (7).
5. A bound electron–hole pair notice rearrangement about ten (7).
7. Scottish dialect with an architectural style (5)?
8. Observe the first bat, a startler (3-6).
9. Identical in many ways but spatially different, is more esoteric perhaps (14).
14. May be high and liable to go off like 15 (9).
16. Switch vessels — strains HP parts (9).
18. Haematite, for example, is a club coin (4, 3).
19. Measure the speed of a bird (7)?
22. Tasteful, spoken about softly (5).
24. Divisions of shocking pests (5).

Crossword no 18

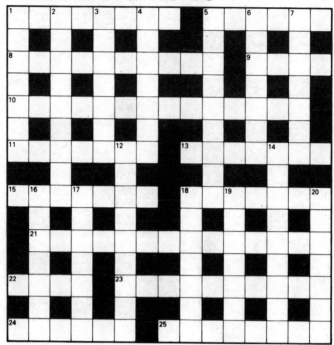

Across

1. There's nothing in the north, the far north, causing sleep (8).
5. Anger about Latin, as a source of Ti (6).
8. Number of articles and steps, one before life begins (10)?
9. Take a quick sleep before the start of the entertainment— a bit of a neck (4)!
10. The coppers almost shone— could have been this (14).
11. Drives away south in splendid confusion leaving North Dakota (7).
13. US petrol container can supply UK cookers (3–4).
15. A shade conservative after what could be the limit (3–4).
18. Madagascar, ideal habitat for a species of worm (7).
21. Scuffle is dumb play with no laughter involved (5–3–5).
22. Horse collar fastener (4).
23. Horse, almost a winner, is an affliction (10).
24. Nothing suppressed, playing our tune out of tune (6).
25. Exceptionally smart appropriating diamond beds (8)

Down

1. Dad appears in celebrated jotter (7).
2. Disastrously in one so unruly, disastrously (9).
3. Couplet rewritten eightfold (7)
4. One gone astray with us like fire (7).
5. We control resistance cooking the roasts (9).
6. Beat a chap for a direction one might fly off at (7).
7. Ella put a note where a swift islander can be found (7).
12. Ludicrous exclamation of repugnance by sailor with the French fore and aft (9).
14. Gymnastic like a rope dancer (9).
16. A tinker can produce the makings of nails (7).
17. An obtrusively ill-bred jumper (7).
18. Postscripts, for example, give more purpose to letter (7).
19. Chest, by the sound of it, suffering from 23 (7)?
20. Prepares gowns (7).

Crossword no 19

Across

4. Of maximum fusibility, summon back with cute internal movements (8).
8. A hundred acres playing hard to get (6).
9. A forming device is *not* an unwelcome guest (8).
10. Three hundred and one in America contrived to produce amber acid (8).
11. Given a little time, fools make something well worth having (6).
12. Firework to adorn a finger (8)?
13. Novel written from an oil rig (8)?
16. Flower, the first of houseboats in yacht disaster (8).
19. She made Popeye's eyes pop (5,3).
21. State of having no remembrance of things past bemused Proust (6).
23. Long lasting, sounds as if you're in the finale (8).
24. Vents for viscounts (8)?
25. Weed to cause annoyance (6).
26. Little devil with metal-bearing veins bursts in (8).

Down

1. Drink, a feature of rugby, some polo and 17 (7).
2. Mixed choir receives bad crit, like the Union Jack (9).
3. Plant has, in a sense, gaseous nitrogen (6).
4. Separation technique derived from choir or steeples (15).
5. Carry in feed for the reversal of pyromellitic acid for example (8).
6. May be across or down, leading to lights (5).
7. Train that is possibly what keeps things moving—or stationary (7).
14. Giant tree. Differentiate it. No, the inverse (9).
15. Find confusion about excellent clue to invisible light (8).
17. I come between Sr and Zr (7).
18. Intimates displayed in glass construction (7).
20. I, too, one might say (6).
22. Page a friend of the bishop (5).

Crossword no 20

Across

1. Authorship of fireworks? It's done with a poker (10).
8. Was Tiberias a wine producer (4)?
10. Reduce in length once more and return to liquidity (10).
11. Anthony's garden (4).
13. Returning a tidal flood in the main, they need free oxygen (7).
15. Moving first to last startles; it's an endearing touch (6).
16. That is to say meanly, with some regeneration (6).
17. Light shed from the animal world (15).
18. Literal meaning: one who rents (6).
20. Chemist famous for a 20 down (6).
21. A look-see, one might say, strange for what is becoming (7).
22. Scant support by the sound of it (4).
25. Nominates plans about tea break (10).
26. Unprepossessing fruit, one hears (4).
27. Printed matter is rubbish from what one hears with nameless character (10).

Down

2. Measure garden (4).
3. Formerly a single time (4).
4. Distances where the deer and the antelope roam (6).
5. Person like Husserl throws gnomes in the pool (15).
6. They tend to agree with society and me in longing (6).
7. Haricot, disposition and coin (6,4).
9. Twice send us into confusion and unexpectedness (10).
12. Rye-based alcohol produced therefore by good man bringing back learning (10).
13. Told positively a travesty of duress (7).
14. Discarded letter cut in pieces by a large number in private room (7).
15. Hue created by youth, inherently a batchelor and depressed (6,4).
19. Disclose article hidden in feast (6).
20. Brook hesitation in one who consumes by fire (6).
23. Restore hat minus offspring (4).
24. Employer employing ruse (4).

Crossword no 21

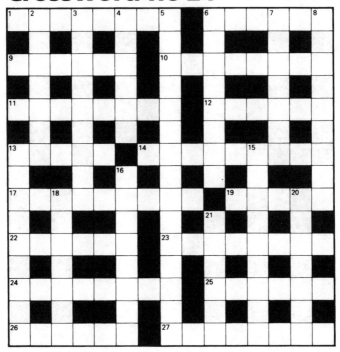

Across

1. Source of wealth or what belongs to me (4–4).
6. Superhuman, though twice slightly drunk in charge (6).
9. Initially any old layabout in family is clay (6).
10. Regret return with narcotic element (8).
11. Stoat with rabbit not a bit distressed by pressure regulator (8).
12. Engineers give up and retreat (6).
13. P in parliament coming back lukewarm (5).
14. Company representative has one good book, or a part (9).
17. Alias Clay, crazy, supplying food (9).
19. Alias Otto, a fragrant essence (5).
22. Plant American uncle in position of bishop (6).
23. Storyteller finds hill on island retreat (8).
24. There's nothing in ugliest parts for the panegyrist (8).
25. Trifles are four or six in four fifths of a set of three (6).
26. Formal offer is painful when touched (6).
27. Smoothing machine and dervish sound like a collection of dates (8).

Down

2. Salt beef source—thereby hangs a strange tale (7).
3. Sad Milton, distressed by colour blindness (9).
4. Light sign—it explodes (6).
5. Clerk Maxwell waves are (15).
6. Public house plot for displaying figures (3–5).
7. I've entered into neat arrangement, achieving natural simplicity (7).
8. Commander, counsellor and one who offers observation (9).
13. Passing troops—first (and last) to entrain in disorder (9).
15. A lousy allowance? It's what's needed to produce TNT (9).
16. Rioter in turmoil inside (8).
18. A medicinal extract got from islands of Langerhans (7).
20. Lively action by English reveals monk's hood (7).
21. Blurt out about a description of what is inhuman (6).

Crossword no 22

Across

1. Each of these is 2.388×10^{-4} 5 down (6).

5. 859.8 5 down h^{-1} (8).

9. 10's theorem places one element after billion in order (8).

10. 100 000 of me on a square metre would make a bar (6).

11. Having stamina and ruining the heart of another new constitution (8).

12. To produce more than one would be the goods (6).

13. A cut strangely indeed strangely is trained (8).

15. Dressed learner taken in by one who is no gentlemen (4).

17. Measure sounds as if it might hurt—take exceptional care (4).

19. Conciliates a tyrant leader in locations (8).

20. Ventilator where Mona Lisa resides (6).

21. Disgusting example of bad language in full new setting (8).

22. Extremely small record (6).

23. Take one variable digit and you're beginning to achieve rigor (8).

24. Fameless, unrecognisable—and identical (4–4).

25. Deceive townsman about fields' centre (6).

Down

2. Trial one organised for, say, a Pekingese (8).

3. Showed the sights to 50 like salts in water (8).

4. Sabbath sanctimonious ceremonial is religious (9).

5. 4186.8 1 across and 1.163×10^{-3} 5 across h (15).

6. Given more than enough of everchanging—dead end (7).

7. Lacking strength in whatever degree at that time in charge (8).

8. A tinge of colour in cutter's eccentricity (8).

14. The east of France extended and cut off (9).

15. As equivalent, abbreviated to see by the sound of it (8).

16. Withering but two thirds manual control (8).

17. Peas hold the means of plant production (8).

18. Repelled and again went head first (8).

19. Very last priest badly taken in by fool's gold (7).

Crossword no 23

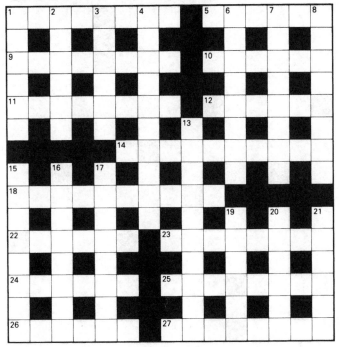

Across

1. Cross worker—he gets around in a purine (8).
5. He has javelin's first broken spear (6).
9. Heather's going around with German, one confining within bounds (8).
10. A mark of distinction conferred, *eg* re-examination outcome (6).
11. Obsessions are recently cultivated blossoms, one hears (8).
12. Security that is given by magistrate (6).
14. String of pearls (50!) found in factory provides fuel for uplift (10).
18. Accomplish disastrous cut, Australia's first in degenerate surroundings (10).
22. Source of heat depicted in Gainsborough's first two exterior paintings (3-3).
23. 500 – *e* (8).
24. Concoction of rice in Pacific (6).
25. Sticky side—have a change (8).
26. Taxer's confounded by additional charges (6).
27. Class try reorganisation for watch glasses (8).

Down

1. Aromatic having three isomers (6).
2. I'm a cloud and I'm a little bit occluded by rising sun (6).
3. Henry at Cox's friend's, a home for bowlers (6).
4. There's no nurse for treatment for such as Al (10).
6. Ready to be won over, so let it be skilful (8).
7. An apparent change in the position of an object (8).
8. Chooses once more, dancing reels *etc* (8).
13. Joiner lets powder explode (10).
15. For each team about which a poem is written a blonde may be so described (8).
16. Iron up smooth waste (8).
17. An ache and its second in Asiatic disorder (8).
19. Emphasise mental pressure (6).
20. Mountain tobacco product in a car crash (6).
21. Holy man stands up to broken line and entrances (6).

Crossword no 24

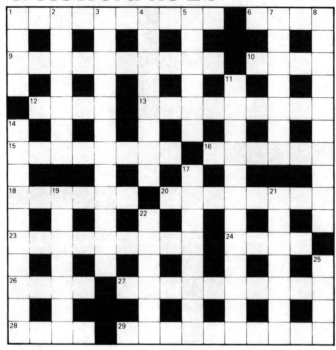

Across

1. Rotatory banking systems, one hears, get in charge (10).
6. Beat about to incite (4).
9. Aphrodisiac can make one violent and not half poorly (4–6).
10. Homeless young Semite? (4).
12. Mix porridge (4).
13. Sugar could be almost valueless, nothing in it (9).
15. Burying organ, an inanimate object (8).
16. A lively, tricky fellow and I are often served with chips (6).
18. A morning with father in Paris, a measure of what gets us lit up (6).
20. Enzyme to break down H_2O_2 and a potassium-free salt cake (8).
23. Moderate medium with its first pair (9).
24. A gem is nothing to a friend (4).
26. Profit from an item of footwear (4).
27. Complex capital reconstruction involving a removal has come about (10).
28. Tranquil command to go gently (4).
29. Length of period in meeting place (10).

Down

1. Bowman's objective or ... (4).
2. Shipyard worker is French and in the stream (7).
3. A good horse might put the internal combustion engine under increased pressure (12).
4. Circumvent ghastly foul tank (8).
5. *Iso-* or *n*-butane, for example – some found in centre of fire (6).
7. Return made by artist in sweep where publican operates (3–4).
8. Differentiate label in ten napkins (5–5).
11. Element, potassium, absorbed in short infrared hot vacuum treatment (12).
14. Lame—at best may be only apparently durable (10).
17. Ocean vessel is a liquid container (5–3).
19. Two postal orders about my first get us solemnly consequential (7).
21. Unyielding first male worker (7).
22. A little maiden in drawers! I'm hooked (6).
25. Yield of one of the best players, by the sound of it (4).

Crossword no 25

Across

1. Veering to port? Not the way to pass it? (13).
10. He sets in motion train at Victoria, leaving one stranded (9).
11. Lines written about undivided American Howler (5).
12. Overgrown, I contended in rivalry (5).
13. Roman accepts wicked lies and recovers from barbarism (9).
14. One objection intended, one hears, for endwise meeting (8).
16. Tarzan's acrobatics—not right lines (6).
19. He is found in Conservative exposition of principle (6).
20. Dolt sounds like an insensitive basket (8).
22. Traveller's loss of half a lb (and gain of other half) results from spinal disorder (9).
24. Smile about a corn (5).
25. Strained time (5).
26. Salutation in pious obscurity (4, 5).
27. Hydrocarbon, for example, mostly came with helicopter crash (13).

Down

2. Assign at payment (9).
3. Six and sixpence for forming brilliant mental images (5).
4. Reconstruction of centre: it is not revealing one's thoughts (8).
5. Do well, making three almost absorb its double up (6).
6. A US battle. Confusion sets in. Orderly columns (9).
7. Take broken guns for steps (5).
8. It is common to Broadcasting House and Chernobyl (13).
9. Property of diamonds: pack of hounds obstruct in it at the end of day (13).
15. Make loose remark about summer for Parisian—he is for the EEC (9).
17. Left a right muddle in New Guinea; it could be a pain in the neck (9).
18. Intimidate, presenting edict to large number, neither north nor south (13).
21. Many heroes, heroines and mistakes are (6).
23. Wash in water in secret (5).
24. Low-down on two spirits (5).

Crossword no 26

Across
1. A young bird, perhaps a notorious Tom and one of what he used (6).
5. Plan the year's first reproduction: heat content (8).
9. If it's natural it's physics (10).
10. Such cats are stinkers (4).
11. Organiser of contest and concert books with hesitation (8).
12. Element briefly expressing surprise (6).
13. One of Hollywood's heavenly bodies (4).
15. Preserve principal town in Scotland (8).
18. Part simple or vulgar possibly (8).
19. Serves without return, oddly in case (4).
21. Obtainable by fission at home; he is not in charge (6).
23. Amphibole formed from arsenic, beryllium and osmium containing bits of catalyst (8).
25. The old hundred lost in captivity (4).
26. I'm cancelled; you might find me in detail (10).
27. Natural at home in this place with books (8).
28. Side is a singular snack (6).

Down
2. There could be a medium (5).
3. Of the nature of nylons and silicones (9).
4. They are second attempts and resist change (6).
5. I conduct trials and live surrounding men (army) in danger (15).
6. 28 in thorny confusion controls metabolism (9).
7. Abundantly in the morning. Make regular journey (5).
8. Produce pills and let one sleep at will (9).
14. One target might well be a producer of malformations (9).
16. Remove to another place to be able to be rendered in another tongue (9).
17. Having discontinuity sounds circumspect (8).
20. Could I be namelessly ignoble? Do me a favour! (6).
22. Fly in amid general confinement ... (5).
24. ... and get fat from return of the cheese board (5).

Crossword no 27

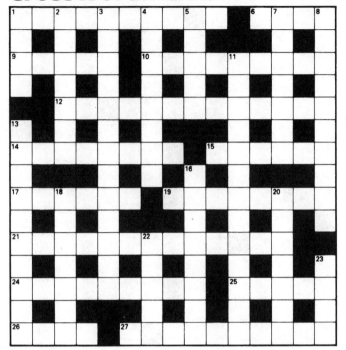

Crossword no 28

Across

1. Mad Monday enables one to switch on (6).
4. Sticking there and lively follower (8).
10. Substances which produce less rot (7).
11. Sanitary principles sound welcome to a man or woman (7).
12. Transverse coils can drive a plane off course (10).
13. Slaughtered lamb can give pain relief (4).
15. Now or then, perhaps, but never never (7).
17. Confounded sailor as he disappeared, first or last (7).
19. Fire or hydrogen, for example, can, with the product of 1 across produce heat (7).
21. Pleases, perhaps, and slips quietly away (7).
23. Liturgy sounds OK (4).
24. Cane chairs can be very sweet (10).
27. Work with a hook – note time lost (7).
28. Homer, for example, here in France, in step, out of step (7).
29. Without their legs bashful spinsters are unscrupulous people (8).
30. The Commander in Chief initially is strangely agitated (6).

Down

1. Wandering ascetic, half dead, is 5 (9).
2. A person of no account in Scotland (7).
3. Smuggler who works only at night? (10).
5. They dread disturbance and 1 down (9).
6. So advances man-eating monster (4).
7. Humbug offering possible ophthalmic benefits (3–4).
8. The marines are useful in reckoning the cost of gas (5).
9. Drink as it arises (4).
14. Staff includes a mother and is to create a smooth surface (10).
16. Dentist's soxhlet (9).
18. Pity! Spiced mixture can make one so (9).
20. Pyrotechnic display lacks 'chic'; it increases this (7).
22. Provider of potential energy for winter sportsmen (3–4).
23. Staggers produced by more than one Bass, for example (5).
25. Reflect: many fell (4).
26. Which? Not this (4).

Crossword no 29

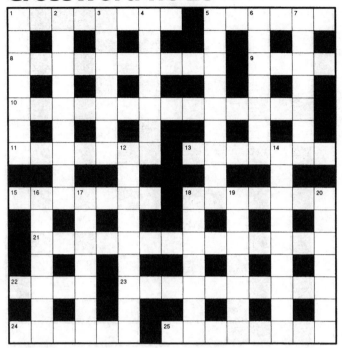

Across

1. Person killed, an occasional labourer and a heartless essay (8).
5. Profit obtained from economic restructuring following company loss (6).
8. Union-approved bombings (3–7)?
9. By the sound of it, a base fiddle (4).
10. Laboratory gas generator invented by H. G. Wells' simple soul (5,9)?
11. Vote in again and dance with electroconvulsive therapy (7).
13. Six footers in schools of opinion (7).
15. State (SE America) accepts nothing, neither solid nor liquid (7).
18. Contact once more for negative improvement (7).
21. A way of determining nitrogen (8,6).
22. Soon wanting a name (4).
23. Tangled cotton thread can give a multifaceted object when the race is lost (10).
24. Members of the 9 family insert nothing into cavities (6).
25. H and D, for example, 'unsoftened' and reorganised opposites (8).

Down

1. Plant often found on a party table (7).
2. They are found on 1 down and in Soho (9).
3. She may be one of 2, or possibly a sitter (7).
4. Mechanical striker to stumble and huff (7).
5. Policy to break inn saucer (9).
6. Examine in quaintly pleasing laboratory receptacle (7).
7. One most upset by Diana's decline (7).
12. Kettles are broken and colours lose nothing (9).
14. Person of spicy character suitable for what might be described as hacking (5–4).
16. Invite to a hop—not a penny awry (7).
17. Endless bird served in topless dinner perhaps (7).
18. Sea lord at sea takes appropriate action after firing (7).
19. A tether is broken—somewhere but not here (7).
20. Thrashings and concealments (7).

Crossword no 30

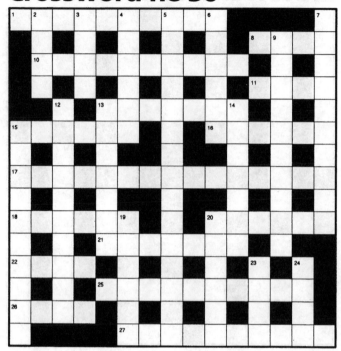

Across

1. Ammeter one might alter to measure air current (10).
8. Shaken for a game of chance (4).
10. Glib revision in recess is open to correction (10).
11. Prevent return of large sums of money (4).
13. Intercede and reflect upon eliminating the end of conflict (7).
15. Gorge that follows tin (6).
16. Special newspaper editions for members of filmed crowds (6).
17. Over half reasonable about nothing to do with the science of widespread outbreaks (15).
18. Imagine a presentation in *Candide* at Edinburgh (6).
20. Washington (and others) sets forth (6).
21. A faithful follower is a company leading light by the sound of it (7).
22. Knacks and wiles (4).
25. Vehicle exhibiting accipitrine skill? (10).
26. Leaderless, dismal deficiency (4).
27. Material offered by prophet to one easily deceived (10).

Down

2. Delightful resort (4).
3. Put me about before, again (4).
4. Miss a clothes horse (6).
5. Could go off and perhaps become a learner or enter tuition (15).
6. Have reference about deceased (6).
7. Clue's disguised in an organised body offering what was once thought to be a bit of light (10).
9. Teach and train, maybe, a provider of heat (10).
12. It's employed in the treatment of heart-sick patients (10).
13. Second article provides more than one impetus (7).
14. Exponent reveals example in former Parisian summer (7).
15. Church mice disturbed by confederate in a way the reader should understand (10).
19. Responses wildly chosen, east rather than north (6).
20. Keeps for sale a means of punishment (6).
23. Leatherhead caught in strange act usually found in the bathroom cupboard (4).
24. Swelling sounds like an eyesore (which it is) (4).

Crossword no 31

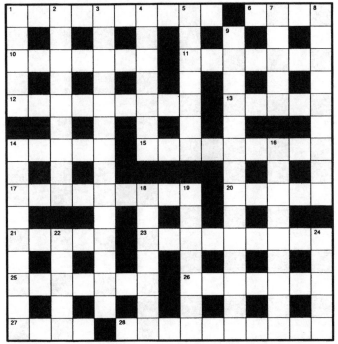

Across

1. A measure of moisture turns mother grey ... (10).

6. ... as one of land does race (4).

10. Storms on the highway are illusory (7).

11. There's salt in tasty paste containing fowl (7).

12. Turns informer encapsulating quiet pub and spells of 9 (9).

13. Strength determined cost—it related to content (5).

14. Metal spearhead removed offensiveness (5).

15. Marriage certificate— projections about advanced state (9).

17. Release craft from what the protectionist prevents (4–5).

20. Employer's right in customs (5).

21. Container attached to metal duct (5).

23. Bird which may be found in disreputable port areas (3–6).

25. To put protective coating on metal is absorbed in rust, for example (7).

26. Cover the limb broken internally (7).

27. Non-U 'W' has its ups and downs (2–2).

28. A cord net is twisted and binding (10).

Down

1. Pong comes to us from decomposed matter (5).

2. 'De-bugger', Monsieur Jean-Paul's here in a state of greed (9).

3. Disgracefully looting caramel compound (14).

4. The inner nature of an alcoholic solution (7).

5. One who makes bare is, perhaps, a retired model (7).

7. Map displays church painting (5).

8. Snack for team winning three points (9).

9. Subjection to rotation may be urgent—and a fiction (14).

14. One escapes an oil plant used for smelling (9).

16. Detestable river crustacean followed by the French (9).

18. Head of abbey abandons book and entertains special constable—an affliction! (7).

19. Extortioner whom, at one time, one would have found in the theatre (7).

22. Half of 27 is mixed up with sin, creating a racket (5).

24. Discuss a source of great gratification (5).

Crossword no 32

Across

4. Course of action gone astray—could be cause of disease (8).

8. British tanker may raise steam (6).

9. Vote Reagan—for a small charge (8).

10. Increasingly cogent source of animal product (8).

11. Stagger back again before wind (6).

12. Relating to mother perhaps and, without hesitation, not intestinal (8).

13. Unequalled as the House of Commons is (8).

16. Allowance including stamp may be rising, or falling (8).

19. To be active I get tidy after a fashion and a degree of wetness (8).

21. Jeers and devours (6).

23. A coiling reappraisal by Wittgenstein perhaps (8).

24. Poison the Heart of Midlothian in nice undulating surroundings (8).

25. Without a penny and become communicative (6).

26. Atomically speaking I'm a proton plus 9 (8).

Down

1. The sound of cabbage and garrulity can keep the home fires burning (4–3).

2. Defamed like Alfred's infamous cakes (9).

3. Vivacious bay carrying King or Queen (6).

4. Purgative indicator (15).

5. Explanations: the alternative is to accept their first (8).

6. I, one hears, could be topic (5).

7. Polymerisation products carelessly expose one (7).

14. Picture composed from Palaces and a disappearance (9).

15. When the first of the English enters, religious abstinence goes by the board! (8).

17. Correct, redress and purify (7).

18. Stretches stocks (7).

20. Male agent might be a source of attraction (6).

22. Student on roof fiddled to earn support (5).

Crossword no 33

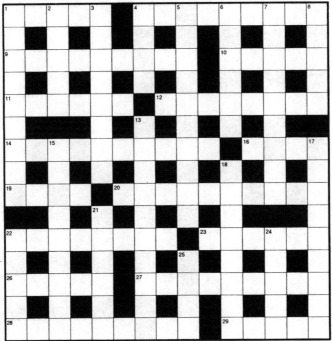

Across

1. Grotesque worker in charge (5).
4. Caught in end—the paper being rewritten after he has left (9).
9. Working well together, ring messy changes (9).
10. For once, a sea change—it's the sea you want (5).
11. Coin that is magnetic, ductile and, briefly, in reverse (6).
12. 14, 18 and 22, for example—a knock-out and learner laid out (8).
14. Novel N_2 chemistry, a mile away from 12 (10).
16. Bank free. Free! (4).
19. Country stain (4).
20. Pause—storm blowing up, creating hazards for Jerry (5–5).
22. Fine face may be revealed by 12 (8).
23. Crude courage can be a source of spirit (6).
26. He drills, the leader of regiment in South African setting (5).
27. An element I am ... (9).
28. ... with shy test mixture we were once thought to have a sobering influence (9).
29. Consumed and corroded (5).

Down

1. SS rations are all criminals need (9).
2. Keynote taken with a gin (5).
3. Money in circulation (8).
4. Revise. It's tied in knots! (4).
5. Seasonableness of imposition wrapped in paper (10).
6. Fragrances when capital is involved, examples of ... (6).
7. ... what are apprehended directly by the senses (9).
8. Perished about noon—partook later (5).
13. Mission can be developed— they are unlikely to sleep on the job (10).
15. Strengthen and control a body of men (9).
17. Those of England go down to the sea in ships (9).
18. A 12, not exactly a protein (8).
21. Scrutinise 10—Chilean vehicle registration (6).
22. Company support—it's deadly (5).
24. Oliver's dance (5).
25. Leg's broken ... and comes together (4).

Crossword no 34

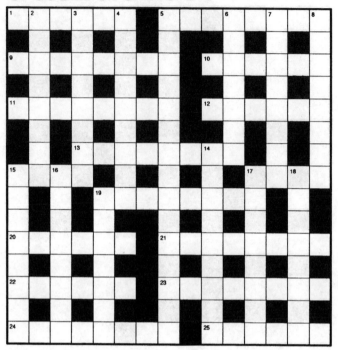

Across

1. Forks and knives change colour (6).
5. Explosive sound of female power (8).
9. Greek in uproar, a nice mess (8).
10. A literary orphan with displeasing odour in a posture (6).
11. Receptacle for 'potions' carefully directed around its centre (8).
12. Weapons displayed by Capone lead to apprehensions (6).
13. PT wear: spotted cat suits? Not quite! PT change needed (8).
15. Nasty eye infection (4).
17. Restrict endless combination of nations (4).
19. We do when we say 'hamadryad' to make a snake angry (8).
20. Lad on the back of a tram shows bottle (6).
21. One who excites a disreputable fool appearing on advanced duty list (8).
22. Prepare again for publication about ebbtide (2–4).
23. Confined in 22 confusion with nine consonants (8).
24. Soldiers ransack the centre of Armenia (8).
25. Is able to betray, one hears, and wipe out (6).

Down

1. Unskilled reconstruction of next pier (8).
3. Flighty mixture of 54 and first five of 13 (8).
4. Managed company, unhappily sour and bitter (9).
5. Develop a condition meant for removal of radioactivity perhaps (15).
6. Goddess in a league's bases (7).
7. Never to be forgotten time in licentious surroundings (8).
8. I'm selfish. Self is an involuntary habitual response (8).
14. Individual is ticket-holder for me: consisting of two (9).
15. Magician famous for his apprentice (8).
16. If you're 8 I'm your prime concern! (8).
17. Stake on a market place—it accelerates the negative (8).
18. Antiquated order about only brief time (8).
19. Lively action associated with English monk's hood (7).

Crossword no 35

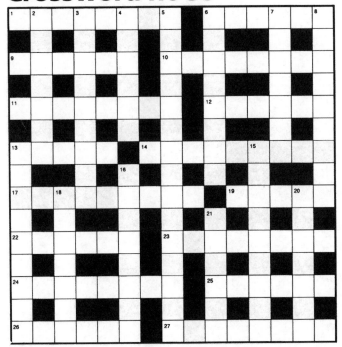

Across

1. Gas needed to cook half slice of Family Pride (8).
6. Write lines about lineage for one who tones up (6).
9. Makes fit girl's set about exercises (6).
10. Made better shape with grass skirt (8).
11. In a way, decay is most frightening (8).
12. Given a flourishing start, this engraver could feed bows (6).
13. Employment sounds as if you make music (5).
14. Odd-ball departing from the centre (9).
17. Bad spirits take in the return of one to rank reductions (9).
19. Disgraces of Brideshead destinies (5).
22. Chance order on ridge (6).
23. Compasses a quiet 26 revolution (8).
24. Heavenly—and others around here (8).
25. Firmly established and poked about (6).
26. Cries unhappily about first of loss and what is left (6).
27. Shouted loudly about retiring warrant officer and went distinctly off-white (8).

Down

2. Exponents reveal large number in what could be east or west (7).
3. Ten operas translated into an auxiliary language (9).
4. Behave in due accordance with all points in artist's return (6).
5. Sanctimonious sounding alternative brought back strictly in a legal sense in the manner of, say, a Catherine wheel (15).
6. Polish, English and Russian are treated to reduce impact (8).
7. Theatrical horse without directions is a 20 (7).
8. Flushed with anger Charlie talks sense initially and aims again (9).
13. Garments submerged by river (9).
15. Taking the long view, worker abandons trunk bearer *in toto* (9).
16. Resolution made by company head (8).
18. A 20 had first in molten mixture (7).
20. Peter mistakenly takes direction for a 7 or 18, for example (7).
21. Descriptive of tributes for all unhappiness (6).

Crossword no 36

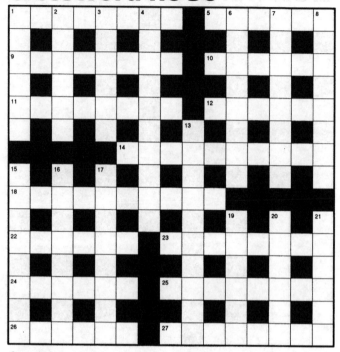

Across

1. Felix has an inclination, one hears, and a means of equilibrating fast (8).
5. Biological 1 (6).
9. I make 1 more efficient and Bruno more wealthy (8).
10. 13 rough riders (6).
11. Prevalent half in impressive surroundings (8).
12. Toss about queer, cheap and nasty material (6).
14. Preliminary sketches from obscene photographs (10).
18. Regular Times people, say, are not English and perished in a bad way (10).
22. Sailor on a cliff is a six-footer (6).
23. Small capitals eraser that takes the s out of gas (8).
24. IOU apt to become more ideal (6).
25. Occupies dresses in taking first place (8).
26. Conclusion: after parking it could be still undecided (6).
27. Incombustible stuff at bosses' disposal (8).

Down

1. Metal peeler (6).
2. Imaginary circle that could be of crab or goat (6).
3. Plunderer is a tailless bird seated on a common toilet (6).
4. Dedicated person is unusually sceptical about, not about (10).
6. No ratings' leader for the navy—it's a cold wind (8).
7. Compliant delivery (8).
8. Such as lamb, say, in bombed sites (8).
13. 10 are second in distressing accidents (10).
15. Urge strongly with almost certain urgency (8).
16. I am deluded—a raindrop splashes, but not the end of the shower (8).
17. Energy quota after a treatment with gas (8).
19. Paralysing extract gives anxiety about our losing our head (6).
20. A music maker, I, clad in battered boots (6).
21. Emergencies provide packs of hounds with inner strength initially (6).

Crossword no 37

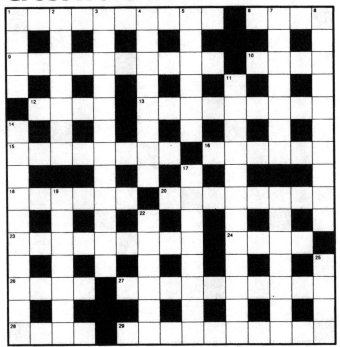

Across

1. One exerting pull and allure takes in copper at centre of rumpus (6,4).
6. It's noble, as none could be or briefly is no less (4).
9. Binding cord in neat knot (10).
10. Worthless people caught in what sounds like an indefinite part (4).
12. Lady with nothing in card dispenser (4).
13. Radon at sixes and sevens— am I at 6? (9).
15. Poe composition about set in place, face to face (8).
16. Falstaff, perhaps, at the end of the day after weekend (6).
18. Poet's written about his last fungi (6).
20. Softening short hairs could be the end of pile (8).
23. CDEISSLXY could be how they see themselves (9).
24. One of thirty which might produce light here (4).
26. Number of legal points of possession could be taken as negative in Frankfurt (4).
27. Eliminating mistakes in notice, grocer confused E and OE initially (10).
28. Mineral sounding like profit, one hears (4).
29. Informal report all points to one who rents (10).

Down

1. Bottle strikes ear as despicable (4).
2. Draw level, going nicely with fish and chips? (5,2).
3. Prepositionally one with truths about second institutions (12).
4. Hue raisers: FBI agents in low places (8).
5. A key handbook (6).
7. Artistic impression of headless drawing (7).
8. Naming bird in marking (10).
11. Ideal of thermal efficiency may yield costly cancer treatment (7,5).
14. Compose melody with numb metal (10).
17. Beds and dances (8).
19. As crazy in scare (7).
21. Specialist, unusually cautious, rejects gold but accepts pound (7).
22. Aerial is spoiled—broken and without tip (6).
25. Jelly, a product of ratis and albat? (4).

Crossword no 38

Across

1. Decay has almost come in sediment (13).
10. Compound derived from any dried mixture with a little heat (9).
11. Stunted Greek character follows commandments, perhaps—a matter of belief (5).
12. Company merger, though only half completed, is one showing promise (5).
13. Pound grew unstable—essential for attempted political coup (9).
14. Irsuteness, jauntiness (8).
16. School Captains' openers entering, fail miserably about revenue (6).
19. They don't get much done, violinists dropping a couple of keys (6).
20. Satisfies what's inside (8).
22. Outward looking, uninhibited character written in poor text (9).
24. 200 mg sounds like an enticement (5).
25. Giving wireless one for nothing, they go from the centre (5).
26. American edentate, peculiar, mad, ill or content (9).
27. Changing Greek character taking time to replace student in motion (13).

Down

2. Fleeting 27 gets me a helper (9).
3. Command fraternity (5).
4. Excessively moralistic renewal of grip is highly topped and tailed (8).
5. Severe and its start are ends (6).
6. Beat first human decorator (9).
7. Admitted name entered was in debt (5).
8. I carry matches unsteadily for sweet determination (13).
9. Strange realisation is about time, a superfluous deprivation of fecundity (13).
15. Telling return managed allowance (9).
17. She issues notes against broken lot (9).
18. Swindle, entice and scorn (8).
21. Be cool and make still (6).
23. Henry was late—perpendicular (5).
24. Boy acted strangely (5).

Crossword no 39

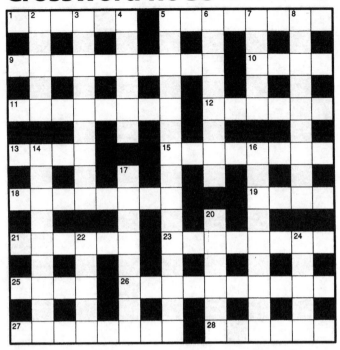

Across

1. Retain parts essential for 9 (6).
5. Ex-coppers in building where merchants meet (8).
9. Tourists' spots, one hears, and the means whereby ... (10).
10. Fly from backhanded smack (4).
11. Weakest grant of land officially approved (8).
12. Charles I tried out acknowledgement (6).
13. National concern in the way (4).
15. Omnipotent, heartless, cruelly disposed, this is suitable for topical application (8).
18. True, one is misrepresented in police return promoting discharge (8).
19. The pith of the matter is in drawing first conclusions (4).
21. Resin used in fabrication of Jubal's amulet (6).
23. Path a boar trots around in alpine retreat (8).
25. Carriage sounds like lubrication for its late occupant (4).
26. Breakdown of union—and it produces a flood (10).
27. Something to be seen through indeed at the end of the war, it is said (8).
28. Tries writing (6).

Down

2. Flower illuminated by a couple of eggheads (5).
3. Antioxidant, for example, made sulphur-free, could become British No 1 (9).
4. Echo solution (6).
5. Choose man for arrest—that's how to create a negative wanderer (8,7).
6. Firm stance: unusual function of 7 (8).
7. Schleswig corner (5)?
8. Ugly Reading street is up, *ie* slopes (9).
14. Handedness of student with tousled hair in London perhaps (9).
16. Rebuilding maybe gets information units (9).
17. Languish about atom-washing compound thought to be toxic (8).
20. Shared by Humber and Wheatstone (6).
22. It's sweet, the sound of a knight on horseback (5).
24. Crazy young 13 comes to crazy conclusion (5).

Crossword no 40

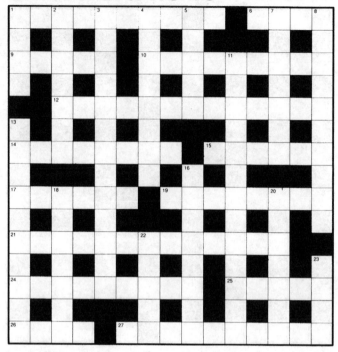

Across

1. Squelched, *ie* treated abominably, without the heart to melt away (10).
6. It cannot be sold to stupefy (4).
9. Dough—a matter of some gravity coming between parents (5).
10. To shed light tried *aria rubato* (9).
12. Nomenclatures tie single room to be demolished (13).
14. Disentangles northern composer in America (8).
15. Gangster's girl takes 10, and the heartless result is descriptive of 9 (6).
17. Ball, one of several in mental faculty (6).
19. Piece of information needed in examinations to reveal where electrons are located (8).
21. Erroneous act; emendation needed to clean up (13).
24. Spirit of wriggling eel unbalanced (9).
25. Gambling, loss of capital leaves enough to decorate the cake (5).
26. Bouquet for an informer (4).
27. They proffer money to the affronted (10).

Down

1. Censure sounds just like mother (4).
2. Large open boat is less heavy—as a result of a spill? (7).
3. Property of carbon court and state drapery (13).
4. Illegible perhaps and desirable leftover (8).
5. The French for proceed reveals the goods (5).
7. Sit uncomfortably under genuine person who is no romantic (7).
8. It's a state of being ingratiating, creatively reassessing its final exception (10).
11. Freshening, I stood on a dire disaster (13).
13. Churl gone astray after dog, almost unable to bark, returns (10).
16. The—*the*—compositions (8).
18. Historic engine, slow starter, climbs rapidly (7).
20. Port in England's earliest basic chemical (7).
22. Heat tarragon extract for essential oil (5).
23. What's laid to start/stop urges (4).

Crossword no 41

Across

1. Like one of the voices in Act IV, Shakespeare's last (6).
4. In a state about 23 down–1 across, 22 (8).
10. Desired time at backward haunt of vice (7).
11. Breast is sharp with elegant extremities (7).
12. Almost moonlit with thick centre undifferentiated throughout (10).
13. Precise kind of light (4).
15. The heartless heart broken and perhaps stirred or bypassed here (7).
17. Fibre revealed by a pack of hounds left in charge (7).
19. Chameleon's heart slowly changes colours (7).
21. Faculty at another time clay (7).
23. They could produce oil—for articulated lorries? (4).
24. He may sell *The Washington Post*, distributing leader to all points (10).
27. Undo 2's work at the station? (7).
28. Weariness with a plump start and a lively dance finish (7).
29. Upset created by balls delivered with spin (8).
30. All a young bird wants, one hears, is the ring (6).

Down

1. My mastery is uneven, showing a lack of balance (9).
2. One under instruction finds article in entire parts assembly (7).
3. Reaction replaces nothing with state in abuse (10).
5. Transmute into something higher: HgCl$_2$ perhaps (9).
6. Chopper chopped is alliance (4).
7. Where writers go for frequent dips (7).
8. Behead guard to make this (5).
9. What producers of leaders for 24s do in ebbtide (4).
14. Itchingly, cautiously, pennilessly surrounds state (10).
16. New Yorker, maybe, a listener imbibing shandy creator (9).
18. Agreeing, one leaves, countering disaster (9).
20. Beneficiary is ambassador to Egyptian capital (7).
22. Get the gen on the saint's ring of truth (7).
23. Broadcast oddly in majority of rhabdoid organs (5).
25. Non-alcoholic firewater first imbibed by drunkard (4).
26. It's silly being inside on a good afternoon (4).

Crossword no 42

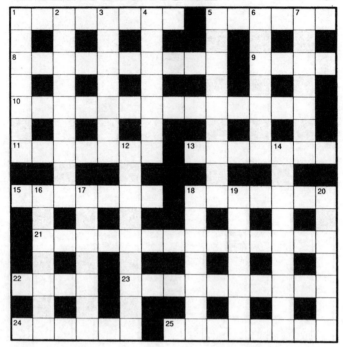

Across

1. Ghastly city mess affecting the body as a whole (8).
5. How rearrangements initially come about for metal (6).
8. Copying by proxy, hearing in-out at sea (10).
9. Displayer of unserviceable plumage in double act (4).
10. Like the study of antiquities accomplished with endless time and reasoning (14).
11. Locks swirl out and percolate back (7).
13. An eccentric lay on the road, probably a 15 (7).
15. Famous walker (or winkler), for example, has hand half in the strong box (7).
18. Hurry and sink coal vessel (7).
21. Deprivations of light seldom experienced by cover girls (14).
22. Render unconscious—crazy coming to (4).
23. Steering gigantic backless van off course (10).
24. Fashionable ladies' standards of perfection (6).
25. Takes for granted and quietly begins again (8).

Down

1. Instrument after being uncapped is standing still (7).
2. Turrets rebuilt with copper insert in arrangement of parts (9).
3. Inter-impress (7).
4. There's nothing in the likenesses; they're post-metamorphic insects (7).
5. Cub leader involved with wild orgy—nice and cool (9).
6. Thorough, fitting tyre round front of car (7).
7. Multiplier of mutating moulds with double centre (7).
12. Awful sinners he puts in a holy place (9).
14. I'm buttery perhaps, and I'm metallic (9).
16. Worried and chased 12 article (7).
17. Virgin needs nothing—and manhandling (7).
18. Undermine breakwater—it's more succulent (7).
19. What an unbroken horse does as tunes are played (7).
20. Flags officers (7).

Crossword no 43

Across

1. Poorly fed men feel he was behind a well stocked table (10).
8. Like Lincoln before the start of day (4).
10. One in the depths receives information at church leading to a parting of the ways (10).
11. Staunch supporter (4).
13. Refer to soldiers—not one out of line (7).
15. Covering of darkness or paleness (6).
16. He infiltrates behind to try once more (6).
17. Appalling R&D cost: searching for dead and alive creature (12,3).
18. Write on edge of manila (6).
20. For control, retain means of publicity, it is said (6).
21. Drawbacks in grooves (7).
22. Club spice (4).
25. Old fashioned and quiet at playing games (10).
26. 'The world's as ..., ay, as sin, and almost as delightful' (4).
27. Beneath that rumble 'erein (10).

Down

2. Bear loses its head in whirlpool (4).
3. Locker owner's lamp (4).
4. Fat queen, a food repository (6).
5. Exerting choice about new agent or bearing minus charge (15).
6. Commission agent, a loud player (6).
7. Foreshadowed a stupid, endless, crazy dearth (10).
9. Flowers and flattery with what the tipsy are in (10).
12. A melancholic, no exception, is a transmutation of the Egyptian art (10).
13. Ignoring copyright, commoner develops plastic precursor (7).
14. Groaners, not primarily artists, can be people like Nat King Cole (7).
15. Transfer oxygen-bearing decomposed organic matter after death (10).
19. Chemist offering comfortable residence for sailors (6).
20. He dreamt up a structure for benzene (6).
23. Fist undoubtedly can produce KO (4).
24. Advantage to move sideways (4).

Crossword no 44

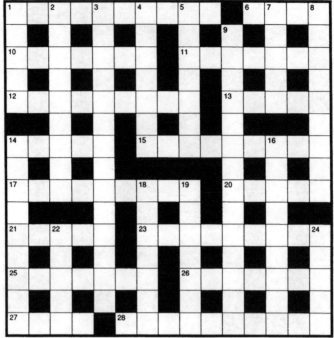

Across

1. You, one hears, must join unruly neophytes to form a line (10).
6. What bellowing Daphnis received from wild, heartless Chloe? (4).
10. Most submissive first of kin is held in poor esteem (7).
11. Make reparation for formerly sanctimonious and converted team leaders (7).
12. Recalls me in what's left of passion spent by sovereign (9).
13. Wanderer is a scholar in the land Cain went to (5).
14. Played back and so on in publicity (5).
15. Shyness—it's confusion building up (9).
17. These vessels should be good pourers (9).
20. Vogues can lead to gear changing at the start of spring (5).
21. Very strong acid derived from mineral oil tails (5).
23. Actress in calamitously shot cases (9).
25. Properly accomplishes results (7).
26. Insulate one likewise coming after usual time (7).
27. Without angle function (4).
28. The guilty Labour leader treasured disturbance (10).

Down

1. Pigeon poet (5).
2. Novel-sounding floor covering, nice but pointlessly air-driven (9).
3. 13, chemistry-crazy; it's part of the chemist's training (14).
4. 24 are crazy about little Margaret (7).
5. Restore native's hat (7).
7. Mad March delight (5).
8. Exaggerates about society's routes to self-destruction (9).
9. Miserable copper sat in his unit counting alpha particles (14).
14. No dangers could be enough to make men men (9).
16. 26 addled Easter egg (9).
18. Enclosed in box with end-wrapping (7).
19. Apricate, then face the time-keeper (7).
22. Small, derisory fine—about a pound (5).
24. They are sown and 16-d in tournaments (5).

Crossword no 45

Across

4. 23 set in place, neither alternative returning (8).
8. Sounds as if you're in fine French department (6).
9. Animal seated on untidy heap, a point far from the sun (8).
10. Intimation ending in itself (8).
11. OMRDAN is such an arrangement (6).
12. 23 from which to get inner out (8).
13. Unthatched one right in an awful hassle (8).
16. Without foundation make happy about sea storm (8).
19. Intellectuals bomb toilet at sea (8).
21. 23s are types of boat (disguised naval ship and its prey), large vessels (6).
23. Piano composition is a hard little piece (8).
24. Dolly Green, an agent used to ensnare (8).
25. Concerned with race and thin ice skating, one gets left behind (6).
26. Carter, for instance, came and trod clumsily (8).

Down

1. Checked—odd sound to badly edit (7).
2. Sounds like two kinds of work when the fighting's over (9).
3. Uncle and I could become central assemblies of 23s (6).
4. Ash generated by unusually sharp person and heat (15).
5. Possibly egg brave chap into swims uphill (8).
6. Cards left after deal and what a bird picks them up with (5).
7. Surd and three arithmetic ciphers absurdly smelly (7).
14. Half of 24 easily digested in the glare of publicity (9).
15. More miserable parts are miles out (8).
17. A coach I've followed is wrong (7).
18. Bucolic assemblage of 500, 150, 100 and 51 × 2 (7).
20. Strangers' heart is turned to stone (6).
22. Set-back in 5 arrangement (5).

Crossword no 46

Across

1. Penniless monarch's company in the great outdoors (8).
6. Beverage abstainer imbibed in carriage (6).
9. She-cat's internal injury (6).
10. Hypnotic of papaveraceous extraction (8).
11. After motorway is enveloped in smoke worker produces wherewithal to clear the air (8).
12. Tape to mend cup-filler (6).
13. Courageous Indian soldier (5).
14. Head actor plays in 'Shapes' (9).
17. Any lasers may be used for polariscope nicols (9).
19. Thieves' talk about cunning about-turn (5).
22. A little bit of heart in a roulade gives extreme pleasure (6).
23. We're often tight and well oiled (8).
24. Torch carried by lover chasing object of affection, though no sweetheart (8).
25. In haste, Fanny Adams embraces one of constant repute (6).
26. Volumes could be written concerning the content of French beds (6).
27. Women lose their head start in departures (8).

Down

2. A film maker and cast leader, Raquel is excited (7).
3. Salvage provides genuine case for rivet repair (9).
4. Understand, understood, up and down, back and forth (6).
5. Moisture process can generate it (7,8).
6. On the bishop's head be they, as 6 across could be (8).
7. Danced very softly, tired out (7).
8. Concerning a candidate coming in like Apollo, say (9).
13. Stunning half of 24, it comes back almost loaded (9).
15. Noble one with head for being ahead of time (9).
16. Staplers can be used to make books of a kind (8).
18. Side by side, sailor's taking a breather by the sound of it (7).
20. A note to Norma, perhaps, would generate work (7).
21. It can provide a clean page, as between sovereigns (6).

Crossword no 47

Across

1. Fish-ing is an industrial process (8).
5. It helps to add at the top of the column (6).
9. Violate prize fighters collectively in fine shemozzle (8).
10. A fellow is quite a dandy (6).
11. Passages taken from past pamphlets (8).
12. Preserve courage (6).
14, Bar Zionist leader writer's
7 firm—it's fixed and it's fundamental (10, 8).
18. To get one over frequency one needs to brandish a piece of cloth (10).
22. Up-front wearing lady's gown leads to imprisonment (6).
23. Judge meatiest stew (8).
24. Fourth prime side (6).
25. Kick a tanner (8).
26. Most recent talks start without least disturbance (6).
27. Remove decomposed reagents (8).

Down

1. Mineral exposed when tree is planted in centre of isle (6).
2. Decadent letter written before holiday (6).
3. British involved in temptation of a group of people (6).
4. They're worn close to the start of term by university members on the Sabbath ... (10).
6. ... where bachelor's more or less muddled about unhinged door (8).
7. See 14.
8. Loitering with intent to get what sounds like cash in intermission (8).
13. The science of forces in equilibrium: oddly it's in data (10).
15. Although it sounds like nonsense it's a measure of gravity (8).
16. Be greedy about hothead and become too hot (8).
17. What sweaty skin does in untidy singlets (8).
19. Cure person who is often giddy, one who naps? (6).
20. Substance of girl in the money (6).
21. Decline in strange creed held by English (6).

Crossword no 48

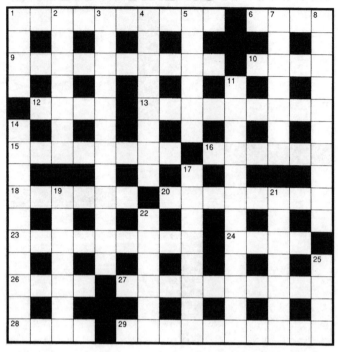

Across

1. Take the explosive out of dynamite—little remains (10).
6. Sediment recovered, one is told, in water in gas trap (4).
9. Coal's other pulverised mineral (10).
10. River and a dam across it (4).
12. Having strength, beer needs its head—but not on top (4).
13. Washing lines about to flutter with no end of a breeze (9).
15. Pirate company thoroughly disorganised like I Pirata (8).
16. Corn in a row, it is said (6).
18. Outlaw also in the arms of a dolly bird (6).
20. Support is only a lock (8).
23. Naughty England team not half involved in complication (9).
24. Intermediate purpose (4).
26. Merely without others worthy to be counted (4).
27. Perceive rise in value (10).
28. Indiscreet record (4).
29. Flowers found in shady parts around mountains (10).

Down

1. Unit difficulty (4).
2. Taste 12 as food should be (7).
3. Acquittals from single allowances (12).
4. Those in the forefront follow girl in charge of electricity (8).
5. Argument has sleazy beginnings (6).
7. It looks bad though they say I'm on the way up (7).
8. Shiny headed (bald) and not a care in the world (10).
11. Her one missing product of protein chemistry that can cure disease (12).
14. Almost wholly distrust dizzy blonde offering a kind of union (6,4).
17. Return of queer doctor about before the time of Crippen, say (8).
19. Provokes the French after all deductions have been made (7).
21. Arise and eat 24 concoction (7).
22. Like some 28s perhaps and wideawake about opposite (6).
25. Catches where Botham practises his art (4).

Crossword no 49

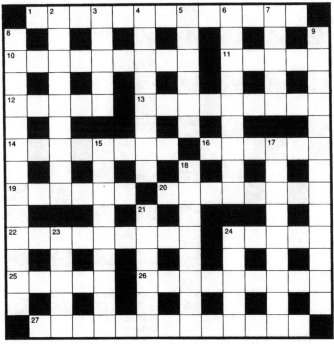

Across

1. It's crude, so prove classy way to make washing soda (6,7).
10. Piece of work not begun by menial taking notice (9).
11. Soothsayer pronounced a bore (5).
12. Of necessity goads the French to withdraw (5).
13. Advocates lie at anchor about interior barrier (9).
14. Inspector offers alternative view first (8).
16. Obscure humour from a numskull (6).
19. Unusually coy about trophy capture (6).
20. Second person is emphatically sane (8).
22. Way below amorous advance (9).
24. Fleece reported as downright (5).
25. Inflamed lump loses head in students' club (5).
26. I'm ready for delivery to you, it's said, without preparation (9).
27. Horrifying dandy died in game (5–8).

Down

2. Harry S. Corbett, one related to midwife (9).
3. It makes us ill, very irritatingly ruining heads for us (5).
4. Annual bakery review contains nothing. Nothing! (4–4).
5. Replay concerning wrong vessel (6).
6. Elucidator who might make ghee from butter (9).
7. Letter from magistrate, half-baked (5).
8. Junior's mould's broken by us—a matter of stress and strain (6,7).
9. Rose featuring unexpectedly as silver glance, for example (13).
15. Dr Zamenhof's brainchild expressed in remarkably neat prose (9).
17. In consequence of which hope we run badly (9).
18. Without recipe I progress badly becoming a source of idle talk (8).
21. Mad woman I accuse of being an inmate (6).
23. Cotton cloth worn in training exercise? (5).
24. Reel in a light boat with LP exchange (5).

Crossword no 50

Across

1. Intuitive feeling about a Shakespearian latter end (6).
5. Fine for the first Windscale (8).
9. Enthusiasts provide conclusive ending to terrible traumas (8).
10. Set high value on mid-west bear (6).
11. A disoriented pointer is an enlarger of the apple of one's eye (8).
12. A goat may get jammed (6).
13. Like a stag's horn extremity in related parts (8).
15. Part of a comb that could be Daniell's (4).
17. Precede plummet (4).
19. Paul's ace played for parts of monkish habits (8).
20. Book concerned with large amounts (6).
21. Centreless circle in front of ruined castle is what stands in the way (8).
22. Go and make good (6).
23. Department of Employment enters to show clearly means of proving disputed fact (8).
24. They disclose former problems (8).
25. Gathers first of artichokes planted in valleys (6).

Down

2. Armour employed in generating power (8).
3. Imaginary return of city at midday: that's about it (8).
4. Cute Irish composition consisting of trial and error (9).
5. Old Queen engineers return to action for conversion (8,7).
6. Flawed report by United Nations (7).
7. Implicit quality of public person (8).
8. Tree bark under a rug upturned (8).
14. Listening could get us engaging in public service (9).
15. Reversed talk (8).
16. Sweetmeat rising typists swallow: what may be sugared to make it palatable (8).
17. He (or she) enables energy production (8).
18. English pointless point of view one can follow (8).
19. More than one top length in suits (7).

Crossword no 51

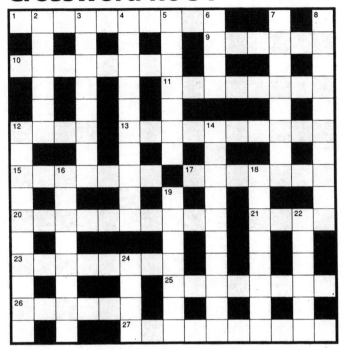

Across

1. Samian implicated in trophy saga (10).
9. Brownish-purple motorway built in stone (6).
10. Logically involved one end, *the* end, in another (8).
11. Perceptions like Bisley targets (8).
12. I'm the 'Z' in ZETA (4).
13. Fungi to notice on growths of shoots (10).
15. Relevant one from Scotland follows first principle (7).
17. Two pets I lost in the tube (7).
20. Marshal goes in rags to establish first act of hostility (10).
21. Inclination to talk in a hypocritical way (4).
23. Plant sprout's about an international unit (8).
25. It's indecent how Charity was said not to behave (8).
26. To goad is unnecessary, doubly pointless (6).
27. Break-up of match is so superficial and divisive (10).

Down

2. Any lousy landing place, it's said, will do for a guy from the West (6).
3. Essential feature of hats found in school? (8).
4. Long tassel, when knotted, can be cause of pain (10).
5. Methyl, for example, is not derived (7).
6. Fool's get-up for resort (4).
7. Decamp in the dark? (5,3).
8. Prettiness *can* be constant (10).
12. Tacking like herringbone (10).
14. Perhaps entomologists can show insects one way inside (10).
16. Highly esteemed evil degrader (8).
18. In full, unshortened *etc* (8).
19. Element could be boron with a trace of iodine (7).
22. Hearts broken by uncle and me subjectively (6).
24. 'Then fall, Caesar!' [Dies]—well almost (4).

Crossword no 52

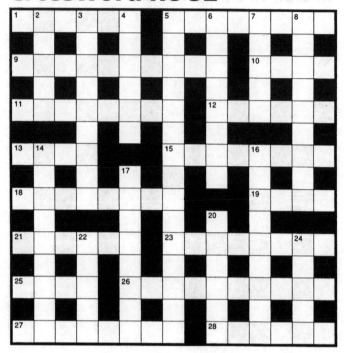

Across

1. Particle mounted minus article (6).
5. Chronicler sounds like a chemist (8).
9. They executed heretics? (10).
10. Legs repair begins to work (4).
11. One created from the first reunion (8).
12. Get away from strife initially into new peace (6).
13. Likewise a loser abandoned by the fleet down under (4).
15. Coordination of chelation not long at rudimentary stage (8).
18. Departments of study, but not the arts (8).
19. Military vehicle sounds as if it might move on, quietly (4).
21. A light riverboat making a racket on the water (6).
23. Trouble arising from display of hesitation in duty (8).
25. Points won in badly presented case (4).
26. Cut a social dash—it's not electrical (10).
27. Fitness from which flows oil or water? (8).
28. Surprisingly the end ended as required ... (6).

Down

2. ... the same end on instrument requiring no adjustment for wash (5).
3. Drunk climbing party on the route between dangerous options (9).
4. Blond type you'll find in a lexicon or dictionary (6).
5. Reassembled sectional board I wickedly employed in quest for gold (8, 7).
6. Coming into being caught on the way up almost in New York (8).
7. Descent of soldier in ascending pass makes sense (5).
8. Sailor with what sounds safe gunpowder component (9).
14. A kind of cell found in prison: closure in each cell's assembly (9).
16. Self-existent lens (9).
17. A clergyman in speech is precise (8).
20. Spirit's in stomach preparation (6).
22. Support offers freedom from difficulty for beginner (5).
24. Weigh plate range (5).

Crossword no 53

Across

1,17. New trading establishment to generate general warming (10, 6).
6. Ring—a greeting left out (4).
9. A little part of the organisation's proficient (5).
10. Gas consumed to cope with BOD (9).
12. Chances for one turned on in responsibilities (13).
14. Gone out of use thus in outside broadcast before summer in Provence (8).
15. Disease wrecks heart before maze (6).
17. See 1 across.
19. Substitute something for Princess's home (8).
21. Speculatively the money is on endless credit to a helper (13).
24. The Devil, Mephistopheles, Satan? (9).
25. Article on 26 words spoken in an undertone (5).
26. Spin half a carcase (4).
27. Compatible sections amended by books (10).

Down

1. Dismal unit of absorbed dose of radiation (4).
2. Frees politician on the way up river having precedence (7).
3. I may go up in pieces—in not a certain glory (13).
4. Workers stand by shattered door; they create a stink (8).
5. Needles seen in a nasty light (5).
7. Covetousness an artist held up in a grip (7).
8. Exaggerates like a flier in America (10).
11. Churchman could be released initially from a special prison (13).
13. They can show the way—it is said lavishly—to supply gems (10).
16. Shows Oxford and Cambridge, say (8).
18. Charged excessively like a billy-goat (7).
20. Ever a startling enmity (7).
22. It's time (but not just 20) an English parliamentarian gets into clothing (5).
23. Trial hosted by Lords (4).

Crossword no 54

Across

1. Sailor, one who leaves a sinking ship, returns an unmanageable chap (6).
4. Criterion of one who is good as well as a rural dean (8).
10. Army chief that might actually be a fast little piece (7).
11. Dark patch on a heavenly body—on the Riviera perhaps (7).
12. Hypothetical stuff ain't cracked up to be of importance (10).
13. Ridiculous departure of sailor—it's irrational (4).
15. Fix a party, one put back for salts (7).
17. Fire or fluorine, say (7).
19. Salt sounds like a witches' sabbath (7).
21. Only deceive in speech expressing deep feeling (7).
23. Said to betray where a hermit lives (4).
24. Data which could be vital (10).
27. Arsonist observed in knotted tie and ring (7).
28. In addition he has other swimming about (7).
29. Hangs us in sheds (8).
30. Fly back east for Paris and Rome (6).

Down

1. Measurement of allowance given to bird (9).
2. Gave details of animal returning with internal spasm (7).
3. Without balance my master is unbalanced in charge (10).
5. Vulgar smack on the French waitress's derrière (9).
6. Muses perhaps? No, says Dieter (4).
7. Container for shot sounds large enough (7).
8. Old-fashioned prosecutor confronted by unruly adolescent (5).
9. Fly raising characteristic smell (4).
14. Sets apart searchers at sea, it is said (10).
16. Afforded refuge from near-perdition in desert storm (9).
18. Shorten means of extending vision (9).
20. Money for special aptitudes (7).
22. Most likely it's fear we need to demolish (7).
23. Winds sounding like locks in Brooklyn (5).
25. Wee snifters for wee people (4).
26. Eyesore encountered in nasty exhibition (4).

Crossword no 55

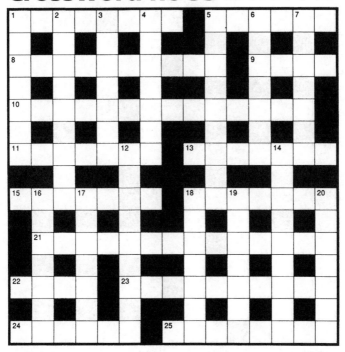

Across

1. See 5.
5,1 Sleeping beauty gave birth to girl in Wonderland, it is said, for 25 (6, 8).
8. Garments we usually gloss over (10).
9. Drink to the Führer's bodyguard (4).
10. Recombustion product divided by π gives neutron excess (8-6).
11. Ignoring lousy start beginners are breadwinners (7).
13. Diabolically masters large quantities coming continuously (7).
15. Vast single unit seen badly (7).
18. Sentimentality that becomes Electra (7).
21. Interesting for a change, fathomless, they result in wholeness again (14).
22. Glint in granite? Copperhead in retrograde design (4).
23. Enlighten politician having time for what sound like whoppers (10).
24. See 25.
25, 5, a display requiring wind
24. and lungs (8,6).

Down

1. Aluminium bearer, a container said to tie badly (7).
2. X-ray photograph may be part of the furniture (9).
3. Halo, gold ring in spinning reel (7).
4. Climbers pocket gold keys (7).
5. As a result, attacker could emerge therefrom (9).
6. Royal Academician panel is capable of being valued (7).
7. Platform or back street hearsay not pertaining to us (7).
12. Comes to rest again in winding street by Latin Quarters (9).
14. Recluse reveals strange reaction about time (9).
16. Cocktail rifle (7).
17. Inscription—mine is an untidy heap (7).
18. Poor Ted gets rotten fish (7).
19. One going in is last of those taken out to storm (7).
20. Nearest could be like, say, Japan (7).

Crossword no 56

Across

1. Series of poems about a dull group of 24 agents (10).
6. Exclusions from slimming diets bring officers back footless (4).
10. Pillar is at the heart of bloke's confusion (7).
11. Affirm period of time required for 3 (7).
12. Trunk bearers break the panels (9).
13. Got out of step about an Argentinian dance (5).
14. Pancake landing of agent within the boundaries of Chile (5).
15. Set English drunk in simple surroundings, thus giving comfort (9).
17. Wearing away one displacing centre of attraction (9).
20. Guard its final departure for arrival on stage (5).
21. Put on alert in time for final Armageddon (5).
23. Teeth shown in heartless sharp reprimands (9).
25. Husbandry shown by money drawer, next to grow old (7).
26. A princess standing still. Fatty! (7).
27. It is said to make higher fish (4).
28. At Scots town, people take first acquisition (10).

Down

1. Throttle hydrogen-bearing fuel (5).
2. Most ingenious climber's first learner on mountain (9).
3. Art of reckoning humble 11 (10,4).
4. Obliquely request to lead off dance (7).
5. Snake from America (mid-west) glides away (7).
7. A profit in return (5).
8. Nosy and petty, perhaps, getting everything down phonetically (9).
9. The subject of Hess's law could make a father notice nothing (4,2,8).
14. Perhaps Hamlet's moral qualities (9).
16. Supple and settled, he's nothing to me (9).
18. Look into what might be a ladybird hosting a prince (7).
19. Blissful state foremost in rainstorm (7).
22. Mix brass maybe (5).
24. Beloved course (5).

Crossword no 57

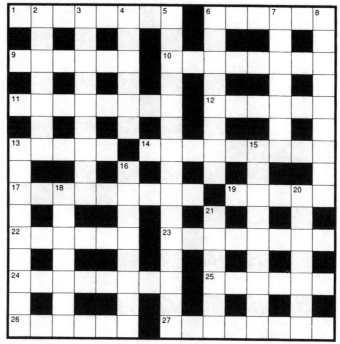

Across

1. Burial place current in, and returning before, grave (8).
6. Nothing comes back in sea-storm like sea-water (6).
9. Smart member of the forest (6).
10. Made a show of preferring to influence journalist (8).
11. Base metal goods from troublesome town (8).
12. Toiler in termitary (6)?
13. Weight: points about fashion (5).
14. Possibly nectarous. Possibly a honey in the palace (9).
17. Diversity of second head (9).
19. Designation sounds like steps useful for those who wish to climb (5).
22. Girl in bed lacks the ability (6).
23. Lead past plant (7).
24. Displeased with bending weapon (8).
25. Fools instinct one has to return south (6).
26. Come to a decision to colonise (6).
27. Managed to corral first brave man, a stock farmer (8).

Down

2. Cleopatra's 'baby', lame, might end up on the roof (7).
3. A call of nine tricks—a cakewalk (9)?
4. Spare nothing in translation for stage works (6).
5. Vehicle cavalry control gives what an engine usefully develops (5,10).
6. Package for 11 (8).
7. Recruits—ie incorporates tank members, on spec initially (7).
8. Suffering—English can under harassment (9).
13. Short pipes opened and stopped by turning keys or handles (9).
15. The bliss a metamorphosis needed for 26 (9).
16. Not reliable without skill, without way (8).
18. 12 away from home is a freebie (7).
20. Boat is as an airship is to air (7).
21. Close union is nothing 16 in a source of amusement (6).

Crossword no 58

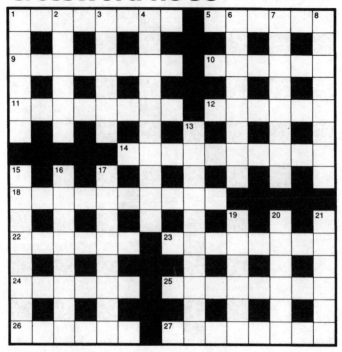

Across

1. It chews the cud before taking a long, bounding stride (8).
5. At times without marbles (6).
9. Standard gone with partners—exemplars (8).
10. Copy element (6).
11. Translate to richer, impressive language (8).
12. Plunders weapons (6).
14. Bobby's depressed with drinking (10).
18. Opposition offered by the Maquis (10).
22. Include element in novel diet to make thinner (6).
23. Dancing Iberians are star couples (8).
24. Steer conversion to sulphur compounds (6).
25. Pernicious mother has an awkward, partial blemish (8).
26. Three times an hour—in a moment (6).
27. Solvent is a blessed mixture containing a little toluene (8).

Down

1. A member before unit (6).
2. Birth pangs of others, maybe (6).
3. Convict has nothing on for a stretch—of water (6).
4. Line placed badly is the substance for line production (6,4).
6. Raving to go out for a hypothetical quantum (8).
7. Arrange a list forbidden near the end (8).
8. Left in, sibling's outside (8).
13. Bring back to life in Shakespeare (act IV) at Elsinore (10).
15. Revise reading at the first inclination (8).
16. Switch to silo—a right stir (8).
17. Coteries can be mysterious (8).
19. Stone is cloth-backed with an openwork fabric (6).
20. Likely subject to an obligation (6).
21. Mistakenly assert, creating bloomers (6).

Crossword no 59

Across

4. Here you'll find Lucifer, a potential partner with whom to spar (8).

8. Force (constable) apprehends maiden with evil ends (6).

9. Bugs and rodents do, inside shop front (8).

10. This zero is OK. Perfect (8)!

11. Haunt of drunken bosses (6).

12. Test unit is crazy—as crazy as you can get (8).

13. Vents cries of one disturbed (8).

16. Kidnap for service of port (8).

19. Contradiction, for example, in people of common descent (8).

21. Invertebrate sounds like strength (6).

23. Impetuosity; eruption coming to a head (8).

24. Deviant: wandering sailor (8).

25. Seaside town on the east side (6).

26. Drinking parties frequented by research chemists (8)?

Down

1, 'Sit up! I'm sober', That's
22. strange—I'm twisted and one-sided (7,5).

2. Assign shares of a penny helping (9).

3. Appeal of a light-headed decoy (6).

4. Resistance to turning when the alarm clock goes off (6,2,7)?

5. Structural pick-me-up for Philip Marlowe (8)?

6. Shoot inaccurately producing jeers of derision (5).

7. Supervise presentation of poetry in Old English (7).

14. Examples in position to point (9).

15. She calls amiss and completely defeats American (8).

17. Strawberry, or old instrument, or symbol in untidy bothy (7).

18. Your uncle's passed a downhill runner (7).

20. Seal blown in anger (6).

22. See 1.

Crossword no 60

Across

1. London, proverbially associated with fire (5).
4. Hide depilatory in 16 tree (9).
9. I am excited about certain quantity of change (9).
10. Bring upon oneself the heart of 28 in dire ruin (5).
11. Trees serving as pillars of the church (6).
12. Art gets confused about one of features (8).
14. Party kiss follows usual beginning and direction—nothing unconventional (10).
16. Convey false impression about voice first recorded in concert (4).
19. Still without wrinkles (4).
20. Smooth, even texture of aniline, say, in 22 down (10).
22. NCO described as violet by Napoleonists (8).
23. Handcart for a heap of earth (6).
26. Strict demand (5).
27. Zulu click is little devil so backed into 16 (9).
28. Groups in formation dances, eights: they prevent relative motion (9).
29. Directions for girl to follow (5).

Down

1. Move in even fashion to be in charge of means of multiplying *etc* (5-4).
2. Humpty-dumptyish poet consumes nothing (5).
3. Damn river vehicle (8).
4. Tobacco money (4).
5. Leaderless Sioux, for example, in perfect setting concerning conception (10).
6. Dub sounds nocturnal (6).
7. Stimulus I have in money first and foremost (9).
8. Aristocrats are leaders of English and Russian (Leningrad) society (5).
13. Mother's employed in travel company for mineral (10).
15. Public, inwardly carefree, offer excessive response (9).
17. Any place but here (9).
18. Apron worn by lass that loved a sailor? (8).
21. Sublime US writer with a twitch (6).
22. Plant Trojan lady without Cretan mountain (5).
24. Attacks of initially rare disease (5).
25. Makes a choice: decapitate church members (4).

Crossword no 61

Across

1,20. Benefit to gardeners is surprisingly not free—huge fees charged initially (10,6).

8. Purpose revealed when a die is cast (4).

10. Just like a man to have a series of books concealed with poor, wicked content (10).

11. Information presented by hand at analytical chemistry seminar (4).

13. After sluggish start sloth returns giving disinclination to act (7).

15. Re-strains tea in earliest of samovars (6).

16. Sewer stitches gown to cap, both backless (6).

17. Repeat resolution of question concerning the process of decision (15).

18. Given a second start, this propulsion would provide upward flight (6).

20. See 1 across.

21. Humour a medium (7).

22. Open a container (4).

25. Displaying panties outside suit should give the papers something to get into (10).

26. Husband and the rest outside are genuine (4).

27. Couturier needs to come into line with God (10).

Down

2. Versed in books, undertook a course of study (4).

3. Provisions held by MP when back to front (4).

4. Birds she gets excited about, and not returns (6).

5. Highest prevailing mood where Appleton, Heaviside and Kennelly were concerned (5,10).

6. Concerning banishment without 50 and 100, each one less (6).

7. Soldier, an unattractive person, shows some susceptibility to Poles (10).

9. Exaggerates importance of caricature missed at Royal Academy (10).

12. More given to bluntness and style. Rubbish! (10).

13. A moment for preparation with little effort (7).

14. Two chaps, identical, with a plant suitable for forage (7).

15. Not a braced construction, fizzy (10).

19. Harnessed—and agreed to differ (6).

20. Puts up the Irish (not English) and the others inside (6).

23. A strange inhospitable area initially—well, some of it is (4).

24. Man perhaps would be ill-advised if accepted by doctor (4).

Crossword no 62

Crossword no 63

Across

1. Festival is a natural disaster (10).
9. Fiddle in an awkward predicament (6).
10,21. Vitamin—given one it could be basic or acidic (8,4).
11. Star possibly the first shellfish to run (8).
12. Encourage a person likely to suceed (4).
13. Eccentric, hurt and surpassed (10).
15. Substance requiring return of regimes, or roughly therein (7).
17. Nothing in diamonds flutter can advance in a winter sea (7).
20. Club shoe that maintains circulation (6,4).
21. See 10.
23. Field work ending times undesirable flows (8).
25. It has me keeled over on wrong side—*ie* listed (8).
26. Mysterious part of Alcazar can end here (6).
27. Voters choose to speak (10).

Down

2. Occupy interest of a bighead thus on advanced lines (6).
3. Partners leave tremendous storm in rainfall measure (8).
4. Vagueness of tailored tiny blouse (10).
5. Do after the usual time about what cows do (7).
6. It's a knird (4).
7. Who follows the dancing sounds like hype (8).
8. Destroyed divine having repeated chaos within (10).
12. 1 across could be like a kangaroo (10).
14. Mechanism for destruction of pest menace (10).
16. The way into delight (8).
18. Shortened undergarment is more showy (8).
19. Joined up to swear about four (7).
22. Purpose diligently applied (6).
24. Song could almost be an elegy (4).

Crossword no 64

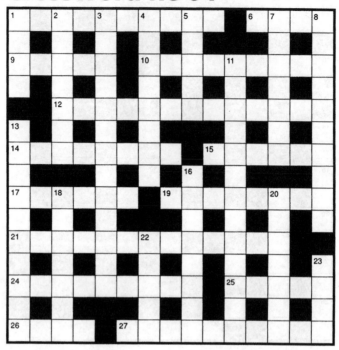

Across
1. Imprudent use of credits in Ethiopian capital (10).
6. Exaggerated military life (4).
9. City for the machine (5).
10. Stones approved in the counties (9).
12. Reach summit of fissure first found around the fireplace (7,6).
14. If you don't start superlatively reliable you get most out of practice (8).
15. There's a mess. Nothing to this place (6).
17. Frightfully alarmed leader jumps out of skin (6).
19. Put another way, reshaper is hoist with his own petard (8).
21. Devotees' ceremony is held in rotisseries (13).
24. From messages sent I always extract something indispensable (9).
25. Fibre bottle (5).
26. River provides supports for drivers (4).
27. Engagements risk rubbish for Hans (10).

Down
1. Working? No; oddly I'm duller (4).
2. Morse, perhaps, without debts. A billion short, this is what he does (7).
3. Requirement fellow in charge found in development of taxonomic groups (13).
4. Impetuosity: a series of unpleasant occurrences going to the head (8).
5. Drained of emotion, yet strangely accepts membership limits (5).
7. Product of 'airdresser and transmitter? (7).
8. Placed charged atom in topside mixture (10).
11. 'Conquer'! Sounds like a cavalry joke (5,8).
13. Derivative of 1 across resembling oil on water (10).
16. Porous like a dungeon (8).
18. Again publish a note for children (7).
20. Country chap emigrates from down under (7).
22. Combine individual part of the whole with its end (5).
23. Sediment in lake nearly met in reverse (4).

Crossword no 65

Across

1. First of many to make 13 across and 16 down 10 across (8).
5. Rock sounds pleasant and precise (6).
8. Motorway police in lager abuse—lethal (10).
9. Cupola provides party for recumbent sweethearts (4).
10. Barge and battered rain hat like 1 across (7,4,3).
11. Cheapens girl as having two cardinal points (7).
13. 10 across like 16 down, but 2 down affected manners with it (7).
15. Weird bloke is a character often found in footnotes (7).
18. Stuck-in-the-mud Gulf leader caught in bed (7).
21. Out of breath in uncommon length: going on a bit too much (4,10).
22. Insect is two thirds materfamilias (4).
23. Poorly, say, once more violating accepted standards in writing (10).
24. Course points to gallows (6).
25. Helmet breaks in crash (8).

Down

1. Unattractive woman with big end went on and got into a financial argument (7).
2. 13 across is rigid, ghastly, unopened book (9).
3. I dined in Works—they're soothing (7).
4. Licentious regimes lead to people living abroad (7).
5. Excite, leaving as shattered (9).
6. Revolutionary ends are causes to love (7).
7. Warrior is a little man found on the old city road (7).
12. Former bizarre Welsh rite (9).
14. Cavalry reverse on which they could be said to be up (9).
16. Merry about everyone, may be filled with 1 across to be 10 across (7).
17. One in a mob set on killing outside the law (7).
18. Near-dead inside the location where a nurse is often found (7).
19. Between June and July diary is confused with the middle of December (7).
20. Make out from scattered cinders (7).

Crossword no 66

Across

1. They moisten and animate the sanctum (10).
6. As a result, maybe, of Anno Domini for example (4).
10. Servitude of bishop on parts of 6 across (7).
11. Next of kin demanding attention in a secluded place (7).
12. Bringing back soldiers' vocation (9).
13. Steps taken by funny sailor at sea (5).
14. Drunk/sober, they are tiresome (5).
15. Distance from stress (9).
17. Delivery from dull-backed partner on lines (9).
20. Journalist covering team members is correct (5).
21. A player disconcerted if thus taken (5).
23. Use a spade—it is about almost all you need for a plant (9).
25. Turps, maybe, is finer (7).
26. Honest, having gained entitlement (7).
27. Carefully inspect container (4).
28. Sense most of length could be a property of elementary particles (10).

Down

1. Given a churn plunger, this process man would be into button and bows (5).
2. Mucked about in charge of maintained hands (9).
3. Writer checks Ireland's dissolution (7,7).
4. Purine base found in port wine (unopened) (7).
5. Three sheets in the wind is about tolerable this evening (7).
7. Glint of medal won in the field—no, the reverse (5).
8. Indeed, procession ended up on the platform (9).
9. Sentimental traveller famous for shandy (8,6).
14. Pretentious tomb, basically friendless, is wrecked (9).
16. Lodge IOU with egghead, confused theorist (9).
18. Approve what sounds like tail-end runner in the National 'unt (7).
19. Rough, rough, heartless try for food (7).
22. They are sharp as interior police (5).
24. Locations said to be good for seers (5).

Crossword no 67

Across

1. It used to occupy a shell; now it could fill one (8).
5. A lover aroused (6).
9. Order doesn't end in unsullied footwork (8).
10. A misfortune to the other side (6).
11. Variable resistance that rose variably (8).
12. In the end all is disrupted in legitimate descent (6).
14. Release from obligation, likewise wash externally (10).
18. Check writer to mark again (10).
22. Cleric breaks a truce (6).
23. Premier who could cover the House ... (8).
24. ... its occupants not beginning to reveal remnants of passion (6).
25. Weather units (8).
26. Directions given to the leader lead astray (6).
27. First letter presses horrendous slanders (8).

Down

1. Let praise float and rise to a great height (6).
2. Comparatively furious source of redness (6).
3. Most pleasant way to a continental resort (6).
4. Parts of breadth are worn out (10).
6. Information about promise to pay is given to strife (8).
7. Losing hair, police get embroiled in drivers' organisation (8).
8. Spartan characters take time about repairing speech defect (8).
13. Blondes, maybe, are unwelcome surprises (10).
15. Biscuits round the bend (8).
16. Undressed brides do in a state of dissipation (8).
17. Coteries may be restricted to an enlightened minority (8).
19. Obstruct motorway in a rare mess (6).
20. Many county songs (6).
21. Takes by force from lively waitress devoid of sloth (6).

Crossword no 68

Across

1. God speaks and confounds (10).
6. Corrode prow off vessel (4).
9. Spreads fertiliser on best garments (3,7).
10. Member to represent in speech (4).
12. Peculiar soft buttocks (4).
13. Subsequently unite on the side (9).
15. Try to catch a singularly useless rattler (8).
16. It's badly placed in lines to ring restaurant (6).
18. Game that is placed between observers (6).
20. Shoeblack accepts a 10 across and sells illicit goods (8).
23. Cover duck in a helpless condition (9).
24. They're drunk in non-teetotal establishments (4).
26. Cordial family dunderhead (4).
27. Put lid back on top of icebox, gradually with care (10).
28. Wine associated with tea and Christmas (4).
29. Once more estimates steps taken by soldiers (10).

Down

1. Go out with a fruit (4).
2. A depressed mental condition—produced by heat (7)?
3. Foreigners accept a measure of acidity from standing passengers (5,7).
4. Sets free sailors about to find a way out (8).
5. Smooth way back for gymkhanas (6).
7. Jaunt allowed for one of three (7).
8. Pursuit on rocky shore is a favourite topic (10).
11. Tricky problems created by improper display of tan and brassiere (5,7).
14. It helps strangers to mix and sea channels to stay open (10).
17. Fabricate and refabricate dice to be identical (8).
19. Places an order for depressions (7).
21. Voter can reveal tolerance with an exception (7).
22. Garment revealed by Boadicea unarticled (6).
25. Easy manoeuvre for those in favour (4).

Crossword no 69

Across
1. Strain wretched trifle (6).
5. Toast double feature (4,4).
9. Containers for fellow to reduce a great deal (10).
10. Long tree (4).
11. Considered after tie knot is repeated (8).
12. Information to put a strain on people (6).
13. Side of paper boy (4).
15. Giving audible expression to absolute end of giving (8).
18. Footballer follows fleet car (8).
19. Audibly grab escorts (4).
21. A major part of sample could be fourth state of matter (6).
23. Charred with the last delivery (8).
25. Live in notice not yet up (4).
26. Marginally upset in a way that causes anxiety (10).
27. Live and dead include way needed for support of those 25 across (8).
28. Is back in, having a large holding, but not coming out on top (6).

Down
2. Oddly it's a lie at a bit of the archipelago (5).
3. Straw diet can be without taste in the extreme (9)
4. It's uncommon to find one in a tarry mess (6).
5. Could possibly bring down bird within the realm of fantasy (5,6,4).
6. Understandings reported in places (8).
7. Bird with its head covered (5)?
8. Wild one caring for lack of knowledge (9).
14. A cover said to be nearly all changed at hand (9).
16. Inmates recover from tiredness (9).
17. Shield from vehicle rapidly (8).
20. Alarmed but not a damaged skin (6).
22. Eleven and 15 spins (5).
24. Surprisingly only put on a new one of pair (5)?

Crossword no 70

Across

1. Selfish in regard to tea drunk. . . (13).
10. . . . or what it's shipped in for artist and players (9).
11. Pretend willingly one hears (5).
12. Lies least uncomfortably (5).
13. Cranks are mistaken for a pillager (9).
14. Develop in island with extremes of temperature (8).
16. Not moving, George maybe has a twitch (6).
19. English nurse can give guarantee (6).
20. Spoke with passion after greeting and got mean (8).
22. Collapse, so I limp on unsteadily (9).
24. Spike has broken point (5).
25. Constable's support to relax beginner (5).
26. What baleen is and who's mixed up with it (9).
27. Starting price altered by one who naps (6,7).

Down

2. Cheek adorned with ornamental fabrics worn by well-dressed ladies (9).
3. Heat treatment devices decarbonise witches (5).
4. Treat us with a ghastly fill (8).
5. Pipes sap (6).
6. Judge, king and thespian can serve stargazer (9).
7. Stupid yokel under the top (5).
8. One across could be adapted for big cases (13).
9. Unusually prudent about English church study, without parallel (13).
15. A physicist, I shall go up after degree followed by seamen at the Open University (9).
17. The grip to develop for use in balancing act (5,4).
18. Leavings controlled by controller (8).
21. VIP emblem or authority for macrocephalic beak (6)?
23. Dad's army food (5).
24. Looks for upper chamber (5).

Solution to crossword no 1

Solution to crossword no 2

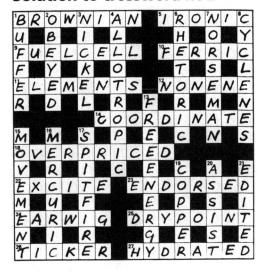

Solution to crossword no 3

Solution to crossword no 4

Solution to crossword no 5

Solution to crossword no 6

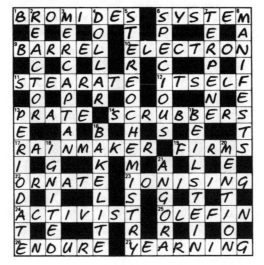

Solution to crossword no 7

Solution to crossword no 8

Solution to crossword no 9

Solution to crossword no 10

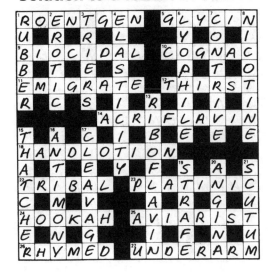

Solution to crossword no 11

Solution to crossword no 12

Solution to crossword no 13

T	H	I	R	T	E	E	N	T	H			G		N
	E		H		P		E		A	L	D	O	S	E
G	A	S	O	L	I	N	E		N			L		E
	V		M		D		D	E	G	R	A	D	E	D
	E		B		I		L		M			M		L
I	N	T	O		O	V	E	R	E	X	C	I	T	E
N			I		R		S		A		N			S
C	O	N	D	U	I	T		P	R	I	M	E	R	S
E		O		T		I		T		A				L
N	I	N	E	T	E	E	N	T	H		I	D	L	Y
D		S				S		Q		L		E		
I	D	E	A	L	I	S	T		U	B	G			
A		N		O			A	N	A	L	O	G	U	E
R	E	S	I	G	N		L		K	A		M		
Y		E			S	E	L	F	E	S	T	E	E	M

Solution to crossword no 14

	E	F	F	E	R	V	E	S	C	E	N	C	E	
A		A		X		O		A		Y		H		T
C	E	N	T	I	G	R	A	M		E	L	A	T	E
R		T		S		T		P		S		R		T
Y	E	A	S	T		E	L	A	S	T	O	M	E	R
L		S				X		N		R		A		
O	X	I	D	I	S	E	R		P	A	R	S	E	C
N		S		S		S		B		I		W		H
I	C	E	B	O	X		B	I	E	N	N	I	A	L
T				T		C		A		V		O		
R	E	C	U	R	S	I	O	N		S	W	E	A	R
I		O		O		T		N		I		L		I
L	A	Y	U	P		R	O	U	N	D	H	E	A	D
E		P		I		U		A		L		Y		E
	Q	U	I	C	K	S	I	L	V	E	R	E	D	

Solution to crossword no 15

Solution to crossword no 16

Solution to crossword no 17

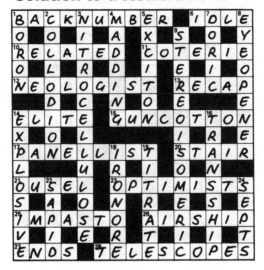

Solution to crossword no 18

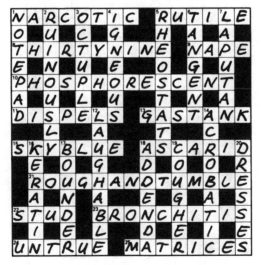

Solution to crossword no 19

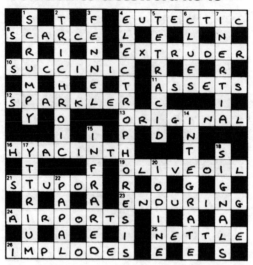

Solution to crossword no 20

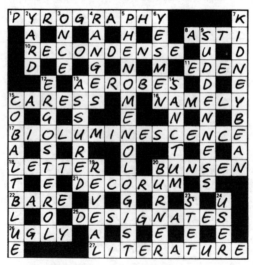

Solution to crossword no 21

¹G	O	³L	⁴D	⁵M	I	N	E	⁶B	I	O	N	I	⁸C

Across and down solution grid reads:

Row 1: G O L D M I N E ■ B I O N I C
Row 2: ■ X ■ A ■ G ■ L ■ A ■ A ■ O
Row 3: ⁹K A O L I N ■ ¹⁰E U R O P I U M
Row 4: ■ L ■ T ■ I ■ C ■ G ■ V ■ M
Row 5: ¹¹B A R O S T A T ■ ¹²R E C E D E
Row 6: ■ T ■ N ■ E ■ R ■ A ■ T ■ N
Row 7: ¹³T E P I D ■ ¹⁴C O M P O N E N T
Row 8: R ■ ■ S ■ ¹⁶I ■ M ■ H ■ I ■ O
Row 9: ¹⁷A L ¹⁸I M E N T A L ■ ¹⁹A T T ²⁰A R
Row 10: N ■ N ■ ■ T ■ G ■ ²¹B ■ R ■ C
Row 11: ²²S E S A M E ■ ²³N A R R A T O R
Row 12: I ■ U ■ ■ R ■ E ■ U ■ T ■ N
Row 13: ²⁴E U L O G I S T ■ ²⁵T R I V I A
Row 14: N ■ I ■ ■ O ■ ■ I ■ A ■ O ■ T
Row 15: ²⁶T E N D E R ■ ²⁷C A L E N D E R

Solution to crossword no 22

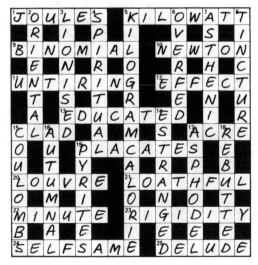

Row 1: ¹J O U L E ⁵S ■ ⁵K I L O W A T T
Row 2: ■ R ■ I ■ P ■ I ■ ■ V ■ S ■ I
Row 3: ⁹B I N O M I A L ■ ¹⁰N E W T O N
Row 4: ■ E ■ N ■ R ■ O ■ ■ R ■ H ■ C
Row 5: ¹¹U N T I R I N G ■ ¹²E F F E C T
Row 6: ■ T ■ S ■ T ■ R ■ ■ E ■ N ■ U
Row 7: ■ A ■ ¹³E D U C A T ¹⁴E D ■ I ■ R
Row 8: ¹⁵C L A D ■ A ■ M ■ S ■ ¹⁷A C R ¹⁸E
Row 9: O ■ U ■ ¹⁹P L A C A T E S ■ E
Row 10: U ■ T ■ Y ■ ■ A ■ R ■ P ■ B
Row 11: ²⁰L O U V R E ■ ²¹L O A T H F U L
Row 12: O ■ M ■ I ■ ■ O ■ N ■ O ■ T
Row 13: ²²M I N U T E ■ ²³R I G I D I T Y
Row 14: B ■ A ■ E ■ ■ I ■ E ■ E ■ E
Row 15: ²⁴S E L F S A M E ■ ²⁵D E L U D E

Solution to crossword no 23

Solution to crossword no 24

Solution to crossword no 25

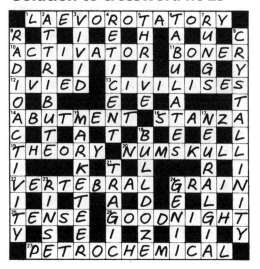

Solution to crossword no 26

Solution to crossword no 27

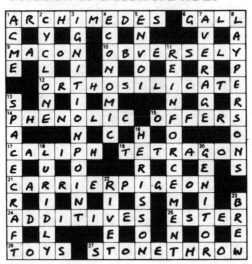

Solution to crossword no 28

Solution to crossword no 29

Solution to crossword no 30

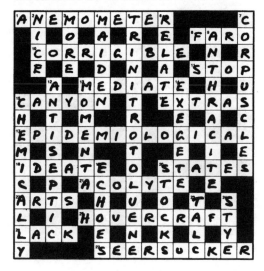

Solution to crossword no 31

Solution to crossword no 32

Solution to crossword no 33

A	N	T	I	C		E	N	T	R	A	P	P	E	D
R		O		U		D		I		R		H		I
S	Y	N	E	R	G	I	S	M		O	C	E	A	N
O		I		R		T		E		M		N		E
N	I	C	K	E	L		A	L	K	A	L	O	I	D
I				N		I		I		S		M		
S	T	R	Y	C	H	N	I	N	E		R	E	E	F
T		E		Y		S		E		A		N		I
S	O	I	L		M	O	U	S	E	T	R	A	P	S
		N		S		M		S		R				H
C	A	F	F	E	I	N	E		B	O	T	T	L	E
O		O		A		I		G		P		W		R
B	O	R	E	R		A	M	E	R	I	C	I	U	M
R		C		C		C		L		N		S		E
A	M	E	T	H	Y	S	T	S		E	A	T	E	N

Solution to crossword no 34

S	I	L	V	E	R		D	Y	N	A	M	I	T	E
	N		O		A		E		L		M		T		G
H	E	L	L	E	N	I	C		A	K	I	M	B	O
	X		A		C		O		A		O			I
S	P	I	T	T	O	O	N		A	L	A	R	M	S
	E		I		R		T		I		T			T
	R		L	E	O	T	A	R	D	S		A		I
S	T	Y	E		U		M		U		B	L	O	C
O		O		A	S	P	I	R	A	T	E		B	
R		U		C			N		L		T		S	
C	A	R	B	O	Y		A	G	I	T	A	T	O	R
E		S		N			T		S		T		L	
R	E	E	D	I	T		I	N	T	E	R	N	E	D
E		L		T			O		I		O		T	
R	I	F	L	E	M	E	N		C	A	N	C	E	L

Solution to crossword no 35

Solution to crossword no 36

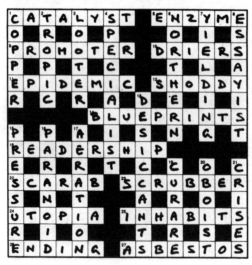

Solution to crossword no 37

Solution to crossword no 38

Solution to crossword no 39

Solution to crossword no 40

Solution to crossword no 41

Solution to crossword no 42

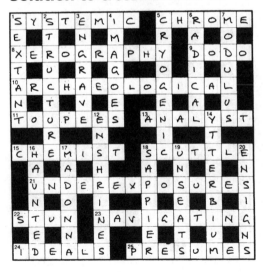

Solution to crossword no 43

Solution to crossword no 44

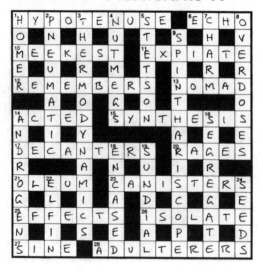

Solution to crossword no 45

Solution to crossword no 46

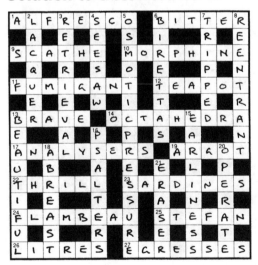

Solution to crossword no 47

Solution to crossword no 48

Solution to crossword no 49

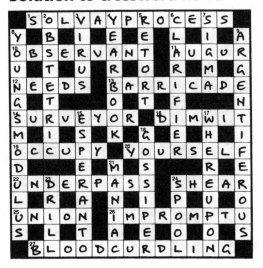

Solution to crossword no 50

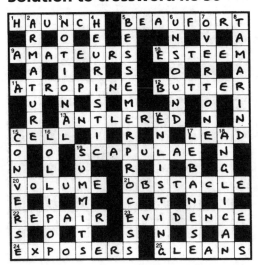

Solution to crossword no 51

Solution to crossword no 52

Solution to crossword no 53

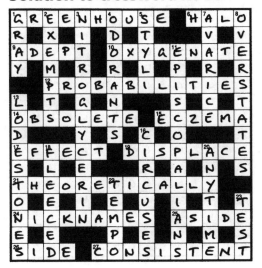

Solution to crossword no 54

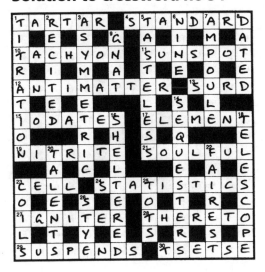

Solution to crossword no 55

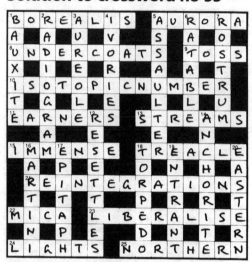

Solution to crossword no 56

Solution to crossword no 57

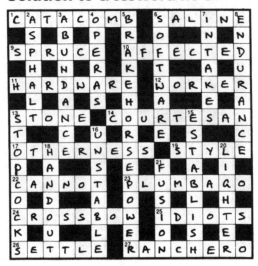

Solution to crossword no 58

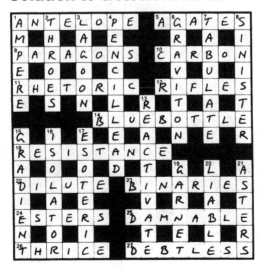

Solution to crossword no 59

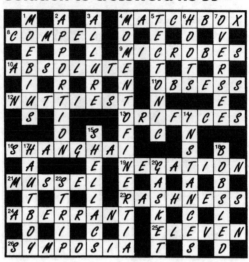

Solution to crossword no 60

Solution to crossword no 61

Solution to crossword no 62

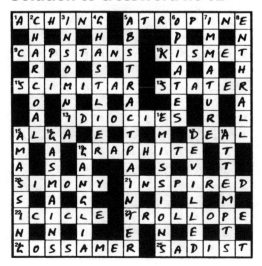

Solution to crossword no 63

Solution to crossword no 64

Solution to crossword no 65

H	Y	D	R	O	G	E	N		G	N	E	I	S	S
A		I		D		M			A		N		A	
G	E	R	M	I	C	I	D	A	L		D	O	M	E
G		I		A		G		V		E		U		
L	I	G	H	T	E	R	T	H	A	N	A	I	R	
E		I		E		E		N		R		A		
D	E	B	A	S	E	S		A	I	R	S	H	I	P
		L		R				S		O		O		
O	B	E	L	I	S	K		B	E	M	I	R	E	D
	A		I		T		E		I		S		A	
	L	O	N	G	W	I	N	D	E	D	N	E	S	S
	L		C		H		S		Y		B		C	
M	O	T	H		I	L	L	I	T	E	R	A	T	E
	O		E		L		D		A		A		E	
E	N	T	R	E	E		B	E	A	R	S	K	I	N

Solution to crossword no 66

H	U	M	E	C	T	A	N	T	S		A	G	E	D
A		A		H		D		O		L		L		E
B	O	N	D	A	G	E		N	E	A	R	E	S	T
E		I		R		N		I		U		A		R
R	E	C	A	L	L	I	N	G		R	U	M	B	A
		U		E		N		H		E		I		
B	O	R	E	S		E	X	T	E	N	S	I	O	N
O		E		D				C		D		E		
M	I	D	W	I	F	E	R	Y		E	M	E	N	D
B			C		N		O		S		O			
A	B	A	C	K		D	I	G	I	T	A	L	I	S
S		C		E		O		H		E		O		I
T	H	I	N	N	E	R		U	P	R	I	G	H	T
I		D		S		S		R		N		U		E
C	A	S	E		G	E	N	T	L	E	N	E	S	S

Solution to crossword no 67

Solution to crossword no 68

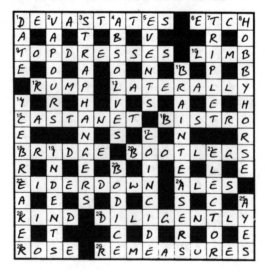

Solution to crossword no 69

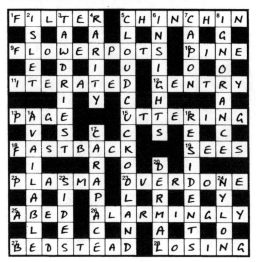

Solution to crossword no 70